SpringerBriefs in Computer Science

Series Editors
Stan Zdonik
Peng Ning
Shashi Shekhar
Jonathan Katz
Xindong Wu
Lakhmi C. Jain
David Padua
Xuemin Shen
Borko Furht
V.S. Subrahmanian
Martial Hebert
Katsushi Ikeuchi
Bruno Siciliano

For further volumes:
http://www.springer.com/series/10028

Yang Liu • Jogesh K. Muppala
Malathi Veeraraghavan • Dong Lin
Mounir Hamdi

Data Center Networks

Topologies, Architectures and Fault-Tolerance Characteristics

 Springer

Yang Liu
Jogesh K. Muppala
Dong Lin
Mounir Hamdi
Department of Computer Science
  and Engineering
The Hong Kong University of Science
  and Technology
Kowloon, Hong Kong, SAR

Malathi Veeraraghavan
Department of Electrical
  and Computer Engineering
University of Virginia
Charlottesville, VA, USA

ISSN 2191-5768   ISSN 2191-5776 (electronic)
ISBN 978-3-319-01948-2   ISBN 978-3-319-01949-9 (eBook)
DOI 10.1007/978-3-319-01949-9
Springer Cham Heidelberg New York Dordrecht London

Library of Congress Control Number: 2013948240

© The Author(s) 2013
This work is subject to copyright. All rights are reserved by the Publisher, whether the whole or part of the material is concerned, specifically the rights of translation, reprinting, reuse of illustrations, recitation, broadcasting, reproduction on microfilms or in any other physical way, and transmission or information storage and retrieval, electronic adaptation, computer software, or by similar or dissimilar methodology now known or hereafter developed. Exempted from this legal reservation are brief excerpts in connection with reviews or scholarly analysis or material supplied specifically for the purpose of being entered and executed on a computer system, for exclusive use by the purchaser of the work. Duplication of this publication or parts thereof is permitted only under the provisions of the Copyright Law of the Publisher's location, in its current version, and permission for use must always be obtained from Springer. Permissions for use may be obtained through RightsLink at the Copyright Clearance Center. Violations are liable to prosecution under the respective Copyright Law.
The use of general descriptive names, registered names, trademarks, service marks, etc. in this publication does not imply, even in the absence of a specific statement, that such names are exempt from the relevant protective laws and regulations and therefore free for general use.
While the advice and information in this book are believed to be true and accurate at the date of publication, neither the authors nor the editors nor the publisher can accept any legal responsibility for any errors or omissions that may be made. The publisher makes no warranty, express or implied, with respect to the material contained herein.

Printed on acid-free paper

Springer is part of Springer Science+Business Media (www.springer.com)

Preface

Large-scale data centers form the core infrastructure support for the ever expanding cloud based services. Thus the performance and dependability characteristics of data centers will have significant impact on the scalability of these services. In particular, the data center network needs to be agile and reconfigurable in order to respond quickly to ever changing application demands and service requirements. Significant research work has been done on designing the data center network topologies in order to improve the performance of data centers.

In this book, we present a detailed overview of data center network architectures and topologies that have appeared in the literature recently. We start with a discussion on various representative data center network topologies, and compare them with respect to several properties in order to highlight their advantages and disadvantages. Thereafter, we discuss several routing algorithms designed for these architectures, and compare them based on various criteria: the basic algorithms to establish connections, the techniques used to gain better performance and the mechanisms for fault-tolerance. A good understanding of the state-of-the-art in data center networks would enable the design of future architectures in order to improve performance and dependability of data centers.

Hong Kong, P. R. China   Yang Liu
Hong Kong, P. R. China   Jogesh K. Muppala
Charlottesville, VA, USA   Malathi Veeraraghavan
Hong Kong, P. R. China   Dong Lin
Hong Kong, P. R. China   Mounir Hamdi

Contents

Acronyms

ABT	Aggregated Bottleneck Throughput
APL	Average Path Length
ARP	Address Resolution Protocol
BFD	Bidirectional Forwarding Detection
BSR	BCube Source Routing
CDN	Component Decomposition Number
DCN	Data Center Networks
DFR	DCell Fault-tolerant Routing
ECMP	Equal Cost Multi-Path
EoR	End of Row
IBA	InfiniBand
IP	Internet Protocol
IPv4/IPv6	Internet Protocol Version 4/6
LCS	Largest Component Size
LDM	Location Discovery Message
LDP	Location Discovery Protocol
LISP	Locator-Identifier Split Protocol
MAC	Medium Access Control
MTBF	Mean Time Between Failures
MTTR	Mean Time To Repair
NIC	Network Interface Card
OSPF	Open Shortest Path First
PBB	Provider Backbone Bridging
RFR	Routing Failure Rate
RSM	Replicated State Machine
SCS	Smallest Component Size
TCP	Transmission Control Protocol
ToR	Top of Rack
TRILL	Transparent Interconnection of Lots of Links
UTP	Unshielded Twisted Pair
VLAN	Virtual Local Area Networks

Chapter 1
Introduction

1.1 Introduction

Data center infrastructure design has recently been receiving significant research interest both from academia and industry, in no small part due to the growing importance of data centers in supporting and sustaining the rapidly growing Cloud-based applications including search (e.g., Google, Bing), video content hosting and distribution (e.g., YouTube, NetFlix), social networking (e.g., Facebook, Twitter), and large-scale computations (e.g., data mining, bioinformatics, indexing). For example, the Microsoft Live online services are supported by a Chicago-based data center, which is one of the largest data centers ever built, spanning more than 700,000 square feet. In particular, cloud computing is characterized as the culmination of the integration of computing and data infrastructures to provide a scalable, agile and cost-effective approach to support the ever-growing critical IT needs (in terms of computation, storage, and applications) of both enterprises and the general public [2, 8].

Massive data centers providing storage form the core of the infrastructure for the Cloud [8]. It is thus imperative that the data center infrastructure, including the data center network, be well designed so that both the deployment and maintenance of the infrastructure is cost-effective. With data availability and security at stake, the role of the data center is more critical than ever.

The topology of the network interconnecting the servers has a significant impact on the agility and reconfigurability of the data center infrastructure to respond to changing application demands and service requirements. Today, data center networks primarily use top of rack (ToR) switches that are interconnected through end of row (EoR) switches, which are in turn connected via core switches. This approach leads to significant bandwidth.oversubscription on the links in the network core [1]. This prompted several researchers to suggest alternate approaches for scalable cost-effective network architectures. According to the reconfigurability of the topology after the deployment of the DCN, there are fixed topology and flexible topology networks. Fixed topology networks can be further classified into

Y. Liu et al., *Data Center Networks*, SpringerBriefs in Computer Science,
DOI 10.1007/978-3-319-01949-9_1, © The Author(s) 2013

two categories: tree-based topologies such as fat-tree[1] and Clos Network [9], and recursive topologies such as DCell [10], BCube [11]. Flexible topologies such as c-Through [14], Helios [7] and OSA [6] enable reconfiguration of their network topology at run time based on the traffic demand. Every approach is characterized by its unique network topology, routing algorithms, fault-tolerance and fault recovery approaches.

Our primary focus in this book is examining data center network topologies that have been proposed in research literature. We start with a discussion on the current state-of-the-art in data center network architectures. Then we examine various representative data center topologies that have been proposed in the research literature, and compare them on several dimensions to highlight the advantages and disadvantages of the topologies. Thereafter, we discuss the routing protocols designed for these topologies, and compare them based on various criteria, such as the algorithms used to gain better performance and the mechanisms for fault-tolerance. Our goal is to bring out the salient features of the different approaches such that these could be used as guidelines in constructing future architectures and routing techniques for data center networks.

We note that other researchers have conducted thorough surveys on other important issues about data center networks, such as routing in data centers [5], and data center virtualization [3]. Kachris et al. published a survey focusing on optical interconnects for data centers [12]. They cover some of the topologies that will be discussed in this paper. Wu et al. made comparisons of some existing DCN architectures [15]. Zhang et al. compared DCN architectures from the perspectives of congestion notification algorithms, TCP incast and power consumption [16].

1.2 Data Center Applications

Cloud computing introduces a paradigm shift in computing where businesses and individuals are no longer required to own and operate dedicated physical computing resources to provide services to their end-users over the Internet. Cloud computing relieves its users from the burdens of provisioning and managing their own data centers and allows them to pay for resources only when they are actually needed and used. Cloud computing can be provided by public clouds (e.g., Amazon EC2, Microsoft Azure) or by private clouds maintained and used within an organization. The services provided by the cloud range from IaaS (infrastructure as a service) where the cloud user requests one or more virtual machines hosted on the cloud resources to SaaS (software as a service) where the cloud user is able to use a service (e.g., Customer Relationship Management software) hosted on cloud resources. The use of cloud computing technology is increasing rapidly and, according to industry estimates, the global cloud computing market is expected to exceed $100 billion by 2015.

Massive data centers hosting the computing, storage, and communication resources form the core of the support infrastructures for cloud computing. With the proliferation of cloud computing and cloud based services, data centers are becoming increasingly large with massive number of servers and massive amount of storage. The entire infrastructure is orchestrated by the Data Center Network to work as a cohesive whole. Server virtualization is increasingly employed to make the data center flexible and adapt to varying demands. In addition, network virtualization is also being considered [3].

Data centers typically run two types of applications: outward facing (e.g., serving web pages to users) applications, and internal computations (e.g., MapReduce for web indexing). In general multiple services run concurrently within a data center, sharing computing and networking resources. Workloads running on a data center are often unpredictable: Demand for new services may spike unexpectedly, and place undue stress on the resources. Furthermore, server failures are to be expected given the large number of servers used [4].

Greenberg et al. [8] define *agility* as an important property, requiring the data center to support running of any service on any server at any time as per the need. This can be supported by turning the servers into a single large fungible pool, thus letting services dynamically expand and contract their footprint as needed. This is already well-supported in terms of both storage and computation, for example by Google's GFS, BigTable, MapReduce etc. This requires the infrastructure to project the illusion to the services of equidistant end-points with non-blocking communication core supporting unlimited workload mobility. The net result is increased service developer productivity, and lower cost while achieving high performance and reliability. In the following section we discuss how these application requirements influence data center network design.

1.3 Data Center Network Requirements

Mysore et al. [13] list the following requirements for scalable, easily manageable, fault-tolerant and efficient Data Center Networks (DCN):

1. Any VM may migrate to any physical machine without the need for a change in its IP address
2. An administrator should not need to configure any switch before deployment
3. Any end host should efficiently communicate with any other end host through any available paths
4. No forwarding loops
5. Failures will be common at this scale, and hence their detection should be rapid and efficient

They then elaborate on the implications of the above requirements on network protocols: a single Layer 2 fabric for entire data center (1&2), MAC forwarding tables with hundreds of thousands entries (3), and efficient routing protocols which disseminate topology changes quickly to all points (5).

The above requirements on the network infrastructure imply that it should provide uniform high capacity between any pair of servers. Furthermore, the capacity between servers must be limited only by their NIC speeds. Topology independence for addition and removal of servers should be supported. Performance isolation should be another requirement, which means that traffic from one application should be unaffected by others. Ease of management supporting "Plug-and-Play" (Layer 2 semantics) should be the norm. The network should support flat addressing, so that any server can have any IP address. Legacy applications depending on broadcast must work unhindered.

1.4 Summary

In this chapter, we presented a basic introduction to data center networks. We briefly mentioned the typical characteristics of data center applications that dictate the expectations from the underlying data center networks. We then briefly reviewed the data center network requirements presented in the literature, which need to be met by future data center networking architectures. The subsequent chapters provide a detailed coverage of various data center network topologies, their performance and fault-tolerance characteristics.

References

1. Al-Fares, M., Loukissas, A., Vahdat, A.: A scalable, commodity data center network architecture. In: Proceedings of the ACM SIGCOMM 2008 Conference on Data Communication, Seattle, pp. 63–74. ACM (2008)
2. Armbrust, M., Fox, A., Griffith, R., Joseph, A.D., Katz, R.H., Konwinski, A., Lee, G., Patterson, D.A., Rabkin, A., Stoica, I., Zaharia, M.: Above the clouds: a berkeley view of cloud computing. Technical Report UCB/EECS-2009-28, EECS Department, University of California, Berkeley (2009). http://www.eecs.berkeley.edu/Pubs/TechRpts/2009/EECS-2009-28.html
3. Bari, M., Boutaba, R., Esteves, R., Granville, L., Podlesny, M., Rabbani, M., Zhang, Q., Zhani, M.: Data center network virtualization: a survey. IEEE Commun. Surv. Tutor. **PP**(99), 1–20 (2012). doi:10.1109/SURV.2012.090512.00043
4. Barroso, L., Hölzle, U.: The datacenter as a computer: an introduction to the design of warehouse-scale machines. Synth. Lect. Comput. Archit. **4**(1), 1–108 (2009)
5. Chen, K., Hu, C., Zhang, X., Zheng, K., Chen, Y., Vasilakos, A.: Survey on routing in data centers: insights and future directions. IEEE Netw. **25**(4), 6–10 (2011)
6. Chen, K., Singla, A., Singh, A., Ramachandran, K., Xu, L., Zhang, Y., Wen, X., Chen, Y.: OSA: an optical switching architecture for data center networks with unprecedented flexibility. In: Proceedings of the 9th USENIX Conference on Networked Systems Design and Implementation, San Jose, pp. 18–18. USENIX Association (2012)
7. Farrington, N., Porter, G., Radhakrishnan, S., Bazzaz, H., Subramanya, V., Fainman, Y., Papen, G., Vahdat, A.: Helios: a hybrid electrical/optical switch architecture for modular data centers. In: ACM SIGCOMM Computer Communication Review, vol. 40, pp. 339–350. ACM, New York, NY, USA (2010)

8. Greenberg, A., Hamilton, J., Maltz, D.A., Patel, P.: The cost of a cloud: research problems in data center networks. SIGCOMM Comput. Commun. Rev. **39**(1), 68–73 (2009). doi:http://doi.acm.org/10.1145/1496091.1496103

9. Greenberg, A., Hamilton, J.R., Jain, N., Kandula, S., Kim, C., Lahiri, P., Maltz, D.A., Patel, P., Sengupta, S.: VL2: a scalable and flexible data center network. SIGCOMM Comput. Commun. Rev. **39**(4), 51–62 (2009). doi:http://doi.acm.org/10.1145/1594977.1592576

10. Guo, C., Wu, H., Tan, K., Shi, L., Zhang, Y., Lu, S.: DCell: a scalable and fault-tolerant network structure for data centers. ACM SIGCOMM Comput. Commun. Rev. **38**(4), 75–86 (2008)

11. Guo, C., Lu, G., Li, D., Wu, H., Zhang, X., Shi, Y., Tian, C., Zhang, Y., Lu, S.: BCube: a high performance, server-centric network architecture for modular data centers. ACM SIGCOMM Comput. Commun. Rev. **39**(4), 63–74 (2009)

12. Kachris, C., Tomkos, I.: A survey on optical interconnects for data centers. IEEE Commun. Surv. Tutor. **14**(4), 1021–1036 (2012). doi:10.1109/SURV.2011.122111.00069

13. Niranjan Mysore, R., Pamboris, A., Farrington, N., Huang, N., Miri, P., Radhakrishnan, S., Subramanya, V., Vahdat, A.: Portland: a scalable fault-tolerant layer 2 data center network fabric. ACM SIGCOMM Comput. Commun. Rev. **39**(4), 39–50 (2009)

14. Wang, G., Andersen, D., Kaminsky, M., Papagiannaki, K., Ng, T., Kozuch, M., Ryan, M.: c-Through: part-time optics in data centers. In: ACM SIGCOMM Computer Communication Review, vol. 40, pp. 327–338. ACM, New York, NY, USA (2010)

15. Wu, K., Xiao, J., Ni, L.: Rethinking the architecture design of data center networks. Front. Comput. Sci. **6**, 596–603 (2012). doi:10.1007/s11704-012-1155-6. http://dx.doi.org/10.1007/s11704-012-1155-6

16. Zhang, Y., Ansari, N.: On architecture design, congestion notification, TCP incast and power consumption in data centers. IEEE Commun. Surv. Tutor. **15**(1), 39–64 (2012)

Chapter 2
Data Center Network Topologies: Current State-of-the-Art

2.1 Typical Data Center Network Topology

A typical data center network configuration is shown in Fig. 2.1. In this configuration, the end hosts connect to top of rack (ToR) switches typically using a 1 GigE link. The ToR switches typically contain 48 GigE ports connecting to the end hosts, and up to four 10 GigE uplinks. The ToR switches sometimes connect to one or more end of row (EoR) switches. The design of the data center network topology is to provide rich connectivity among the ToR switches so that the requirements set out in Sect. 1.3 are satisfied.

Forwarding of packets may be handled either at Layer 3 or Layer 2 depending on the architecture. If the Layer 3 approach is used, then IP addresses are assigned to hosts hierarchically based on their directly connected switch. Standard intra-domain routing protocols, eg. OSPF, can be used for routing. However this approach incurs large administration overhead. An alternative is to use a Layer 2 approach. Here address assignment is based on flat MAC addresses and forwarding is done accordingly. This approach incurs less administrative overhead. However this approach suffers from poor scalability and low performance. A middle ground between Layer 2 and Layer 3 is to use virtual LANs (VLANs). This approach is feasible for smaller scale topologies, however it suffers from the resource partitioning problem [5].

Two types of traffic shown in Fig. 2.1, viz., North-South traffic and East-West traffic, place different demands on the networking infrastructure. North-South traffic corresponds to communication between the servers and the external world. East-West traffic is the internal communication among the servers. Depending on the type of application (outward facing or internal computation), one or the other type of traffic is dominant.

End host virtualization is extensively used in today's data centers, enabling physical servers to support multiple virtual machines (VM). One consequence of this approach is the need to support a large number of addresses and VM migrations (e.g. vMotion). In a layer 3 fabric, migrating a VM to a different switch changes

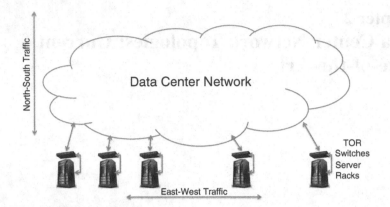

Fig. 2.1 A typical data center network topology

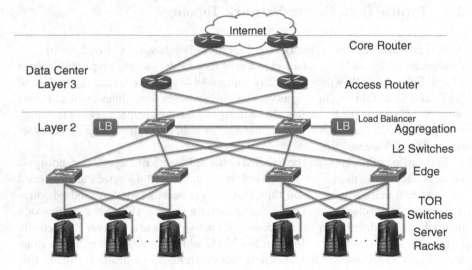

Fig. 2.2 Cisco's recommended DCN topology

the VM's IP address. In a layer 2 fabric, migrating a VM incurs ARP overhead, and requires forwarding on millions of flat MAC addresses.

2.1.1 Tree-Based Topology

Tree-based topologies have been the mainstay of data center networks. As an example, Cisco [2] recommends a multi-tier tree-based topology as shown in Fig. 2.2. ToR switches connect to edge switches, and edge switches in turn connect to aggregation switches. Aggregation switches in turn are connected to the core.

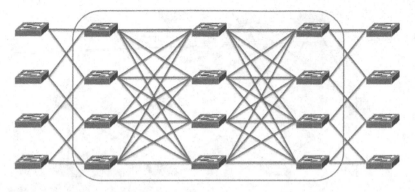

Fig. 2.3 A 5-stage Clos network topology

The network is designed to be hierarchical with $1 + 1$ redundancy to deal with failures. Switches higher up in the hierarchy handle more traffic, which necessitates fault-tolerance features to ensure availability. This topology allows for scalability.

Aggregation switches define the boundary for a Layer 2 domain. In order to contain the broadcast traffic (e.g., ARP), this domain is further subdivided into VLANs.

2.1.2 Clos Network

A Clos network [3] was originally designed as a multi-stage circuit switched interconnection network to provide a scalable approach to build large scale voice switches. A typical Clos network consists of three levels of switches are termed *input*, *middle*, and *output* switches in the original Clos terminology. As an example, a 5-stage Clos network topology is illustrated in Fig. 2.3. This architecture provides a statistically non-blocking core around which the network infrastructure can be built.

Clos network based interconnection has seen increasing deployment as a data center network infrastructure [14]. When used in data center infrastructures, the Clos network is folded around the middle to form a folded-Clos [3] network as depicted in Fig. 2.4. This is also referred to as a leaf-spine topology, and can also be viewed as a multi-rooted tree. When deployed in data center networks, the network usually consists of three levels of switches: the ToR switches directly connected to servers, the aggregation switches connected to the ToR switches, and the intermediate switches connected to the aggregation switches.

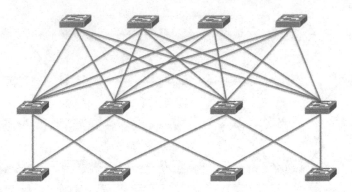

Fig. 2.4 A leaf-spine network topology

2.2 Data Center Network Technologies

Two network fabrics commonly used in data center networks are Ethernet and
InfiniBand (IBA) [9]. Another high-speed network technology is Myrinet [1]. In this
book, we primarily focus on data center networks that are based on Ethernet.
As such, InfiniBand and Myrinet are outside the scope of this book. Interested
readers may find more information about these technologies and their use in data
centers in [9].

Ethernet is commonly used in data center networks. The nodes can be configured
to operate in Ethernet-switched mode or IP-routed mode. Pros and cons of these two
modes of operation are discussed below. Scalability is especially a concern since
data center networks increasingly connect 100K–1 million servers. Characteristics
for comparison include forwarding table size, address configuration, VM migration,
and routing efficiency.

The Ethernet 6-byte MAC addressing is flat, i.e., an interface can be assigned
any address (typically by the manufacturer) without consideration of its topological
location (i.e., the switch/router to which it is connected). On the other hand,
IP addressing is hierarchical, which means the address assigned to an interface
depends on its topological location. The advantage of hierarchical addressing is
that addresses can be aggregated for use in forwarding tables, which makes the
size of forwarding tables smaller than in networks with flat addressing. This makes
IP-routed networks more scalable than Ethernet-switched networks, and explains
why the Internet, which interconnects multiple networks, each owned and operated
by a different organization, uses IP routing.

Networks with flat addressing, e.g., Ethernet-switched networks, require no
address configuration. Server interfaces come ready for plug-and-play deployment
with manufacturer-configured addresses. With any hierarchical addressing scheme,
address configuration is required since addresses have to be assigned to server
interfaces based on their topological locations. While this step can be automated,
it is still a source of potential misconfiguration errors [11].

With respect to *VM migration*, the IP-routed mode has a disadvantage. To handle VM migration in IP-routed networks, one of three approaches could be used. In the *first* approach, the migrating VM's IP address is changed to reflect its new topological position, allowing for reachability to be ensured because its new address will be part of a summarized address that is advertised by the routing protocol. However, any ongoing TCP connections will be lost when the VM is migrated, because unlike SCTP [13], TCP does not support dynamic address configuration. In the *second* approach, forwarding tables in all intermediate routers are updated with the /32 IPv4 (and /128 IPv6) address of the migrating machine; this is an inherently unscalable solution. In the third approach, a solution such as mobile IP [15] is used in which home agents (routers) register foreign care-of addresses for the migrating machine, and tunnel packets to the migrating machine at its new location. In contrast, VM migration in an Ethernet-switched network is simpler. The migrating machine can simply retain its IP address; what needs to change is the IP-address-to-MAC-address resolution stored at the senders. One solution to this problem noted in an IBM VM migration document [7] is as follows: "Once traffic from the Virtual Machine on the new node is sent again the ARP tables will get updated and communication will continue as expected."

Finally consider *routing efficiency*. Routing protocols, used to disseminate reachability information, are less feasible in networks with flat addressing. For example, Ethernet-switched networks have no explicit routing protocols to spread reachability information about the flat addresses of server NICs. Instead flooding and address learning are the methods used to slowly create forwarding table entries. Flooding brings in the specter of routing loops, which necessitates the spanning tree protocol. The spanning tree protocol avoids loops by disabling certain links, but this is wasteful of network bandwidth resources. Modified versions of the spanning tree, e.g., multiple spanning trees using VLANs, have been standardized in IEEE 802.1Q [8] to address some of these issues. However, Kim et al. [10] note some drawbacks of using VLANs such as configuration overhead, limited control-plane scalability, and insufficient data-plane efficiency. IP routing protocols such as OSPF are often augmented with Equal Cost MultiPath (ECMP) techniques [17] to enable the network to use parallel links between two nodes without needing to disable one as with the spanning tree protocol. With ECMP, packet headers are hashed before one of multiple paths is selected to avoid out-of-sequence packets within individual flows. Therefore, IP-routed networks have the distinct advantage of more efficient routing when compared to Ethernet-switched networks.

Given the above considerations, it appears that neither a completely flat addressing network (e.g., Ethernet-switched network) nor a completely hierarchical addressing network (e.g., IP-routed network) is ideal to meet the requirements of scalability (small forwarding tables), ease of address configuration, VM migration, and routing efficiency. Interestingly, these features are required in both data center networks and in the global Internet. Solutions, which are hybrid approaches combining both flat and hierarchical addressing schemes, proposed for both arenas are remarkably similar. In the wide-area, Locator-Identifier Split Protocol (LISP) [4] is a representative of a set of solutions that use hierarchical addresses as locators

in the core, and flat addresses as identifiers in the edges. Equivalent approaches in
the data center network arena are IETF's TRansparent Interconnection of Lots of
Links (TRILL) [16] and the Provider Backbone Bridging (PBB) approach in IEEE's
Shortest Path Bridging solution [8]. Tunneling is used in that the original packets
are encapsulated in new packets whose outer headers carry the addresses used for
packet forwarding through the core (e.g., IP-in-IP, TRILL or MAC-in-MAC). One
can visualize this solution as a network-in-network, with the address space used in
the inner core network being different from that used in the outer network. Since
both the core and the outer networks are connectionless packet-switched networks,
the network-layer header[1] is repeated, one for forwarding within the core, and one
for forwarding at the edges. In a second set of solutions, tunneling is extended all
the way to the servers. This approach is proposed in shim6 [12] for the Internet, and
in VL2 [6] for data center networks.

2.3 Problems and Issues with Current Approaches

Whether or not these standards based solutions, such as TRILL and PBB, are
sufficient to meet the requirements of data center networks is not explored by most
research publications. Instead researchers highlight problems related to scalability,
throughput, data integrity and system availability, power efficiency, and operational
costs. For example, consider the problems described in Greenberg et al. [5] and
Zhang et al. [18].

*   *Scalability.* Besides the capability of hosting a large number of servers, DCNs
    also need the ability of easily scale in situ to meet future demands. The current
    topologies, however, do not offer such scalability. The only way is to scale up
    requiring a total replacement of switching devices and rewiring. A scale out
    approach is instead preferred for future topologies.
*   *Poor Server-to-Server Connectivity.* The hierarchical design of conventional
    DCN topologies makes the upper levels highly oversubscribed, leading to high
    probability of congestion. This is especially severe in tree-based topologies. As a
    result, most of the bandwidth towards the edge of the network will be in fact
    wasted and the average server-to-server bandwidth will be poor.
*   *Static Resource Assignment.* Conventional approaches employ VLAN
    technology to split the DCN to satisfy special requirements of individual
    applications. VLAN approach offers some performance and security isolation,
    but it has disadvantages considering agility: VLAN integrates multiple aspects
    such as traffic management, security and performance isolation, and the traffic
    of VLAN requires high oversubscription in the network core.

---

[1]Ethernet is considered Layer 2, though switching is defined in the OSI protocol reference model
as a network-layer function.

- *Resource Fragmentation*. Load balancing in conventional DCN designs limits the resources that an application can get when its requirement grows. This in fact splits the DCN into small subnets, and may restrict some applications while wasting resources at the same time.
- *Fault-Tolerance*. The design of conventional DCN topologies makes them vulnerable to failures in the network core, which is inevitable considering the scale of DCNs. Besides, conventional designs do not offer mechanisms for dealing properly with network failures in a manner that reduces their impact.

2.4 Summary

Current state-of-the-art data center networks are typically built using hierarchical basic-tree and multi-rooted tree topologies. In this chapter, we reviewed some of the salient features of these topologies. We also discussed how Ethernet is used in current data center networks. After describing the pros and cons of Ethernet-switched vs. IP-routed networks, we mentioned new solutions such as TRILL and PBB that are designed to overcome drawbacks with both flat addressing and hierarchical addressing schemes.

References

1. Boden, N., Cohen, D., Felderman, R., Kulawik, A., Seitz, C., Seizovic, J., Su, W.: Myrinet: a gigabit-per-second local area network. IEEE Micro **15**(1), 29–36 (1995)
2. Cisco: Cisco data center infrastructure 2.5 design guide. http://www.cisco.com/en/US/docs/solutions/Enterprise/Data_Center/DC_Infra2_5/DCI_SRND_2_5a_book.html
3. Dally, W., Towles, B.: Principles and Practices of Interconnection Networks. Morgan Kaufmann, Amsterdam/San Francisco (2004)
4. Farinacci, D., Fuller, V., Meyer, D., Lewis, D.: Locator/ID separation protocol (LISP). Internet Draft (draft-ietf-lisp-23) (2012). (Work in progress)
5. Greenberg, A., Hamilton, J., Maltz, D.A., Patel, P.: The cost of a cloud: research problems in data center networks. SIGCOMM Comput. Commun. Rev. **39**(1), 68–73 (2009). doi:http://doi.acm.org/10.1145/1496091.1496103
6. Greenberg, A., Hamilton, J.R., Jain, N., Kandula, S., Kim, C., Lahiri, P., Maltz, D.A., Patel, P., Sengupta, S.: VL2: a scalable and flexible data center network. SIGCOMM Comput. Commun. Rev. **39**(4), 51–62 (2009). doi:http://doi.acm.org/10.1145/1594977.1592576
7. Hyper-v quick migration basp team connectivity loss – ibm bladecenter and system x. http://www-947.ibm.com/support/entry/portal/docdisplay?lndocid=migr-5087137
8. IEEE Standards Association: Media access control (mac) bridges and virtual bridge local area networks (2011)
9. Kant, K.: Data center evolution: a tutorial on state of the art, issues, and challenges. Comput. Netw. **53**(17), 2939–2965 (2009)
10. Kim, C., Caesar, M., Rexford, J.: Floodless in seattle: a scalable ethernet architecture for large enterprises. ACM SIGCOMM Comput. Commun. Rev. **38**(4), 3–14 (2008)

11. Ma, X., Hu, C., Chen, K., Zhang, C., Zhang, H., Zheng, K., Chen, Y., Sun, X.: Error tolerant address configuration for data center networks with malfunctioning devices. In: 2012 IEEE 32nd International Conference on Distributed Computing Systems (ICDCS), Macau, pp. 708–717. IEEE (2012)
12. Nordmark, E., Bagnulo, M.: Shim6: level 3 multihoming Shim protocol for IPv6. RFC 5533 (Proposed Standard) (2009). http://www.ietf.org/rfc/rfc5533.txt
13. Ong, L., Yoakum, J.: An introduction to the stream control transmission protocol (SCTP). RFC 3286 (Informational) (2002). http://www.ietf.org/rfc/rfc3286.txt
14. Pepelnjak, I.: Data center fabrics – what really matters. http://demo.ipspace.net/get/Data%20Center%20Fabrics%20What%20Really%20Matters%20RIPE.pdf
15. Perkins, C.: IP mobility support. RFC 2002 (Proposed Standard) (1996). http://www.ietf.org/rfc/rfc2002.txt. Obsoleted by RFC 3220, updated by RFC 2290
16. Perlman, R., Eastlake 3rd, D., Dutt, D., Gai, S., Ghanwani, A.: Routing bridges (RBridges): base protocol specification. RFC 6325 (Proposed Standard) (2011). http://www.ietf.org/rfc/rfc6325.txt. Updated by RFCs 6327, 6439
17. Thaler, D., Hopps, C.: Multipath issues in unicast and multicast next-hop selection. RFC 2991 (Informational) (2000). http://www.ietf.org/rfc/rfc2991.txt
18. Zhang, Y., Ansari, N.: On architecture design, congestion notification, TCP incast and power consumption in data centers. IEEE Commun. Surv. Tutor. 15(1), 39-64 (2012)

Chapter 3
Data Center Network Topologies: Research Proposals

3.1 Classification of Topologies

In this book, we choose to primarily concentrate on a representative subset of data center network architectures that have appeared in the literature. We classify these architectures as fixed-topology architectures and flexible-topology architectures, according to whether the network topology is fixed from the time it is deployed. Fixed-topology architectures based on (i) Fat-trees include the architecture proposed by Al-Fares et al. [2], PortLand [18] and Hedera architectures [3], (ii) Clos networks as represented by the VL2 architecture [11], and (iii) Recursive topologies such as the DCell [12] and BCube architectures [13]. Flexible-topology architectures include c-Through[20], Helios [9] and OSA [5]. Other noteworthy architectures that we do not explicitly cover in this book, but are worth noting include FiConn [15], MDCube [21], and CamCube [1]. Figure 3.1 gives a taxonomy of the different data center network topologies.

The next few chapters describe this set of data center network architectures all of which were proposed in research literature. In most of these architectures, the starting point is a specific topology (Chap. 3). By tying the addressing scheme to the topology, a routing protocol is no longer needed. Instead each node can run a predetermined algorithm to determine the output link on which to forward a packet based on its destination address (Chap. 4). Multipath routing and other techniques are often included to improve performance (Chap. 5). Also, as the regular structure of the topology will break down when failures occur, automated procedures are required to update forwarding tables (Chap. 6).

Furthermore, here we summarise the notation used in this section:

- n: The number of ports on a switch.
- k: The number of ports on a server.
- N: The total number of servers inside a data center network.

Y. Liu et al., *Data Center Networks*, SpringerBriefs in Computer Science, DOI 10.1007/978-3-319-01949-9_3, © The Author(s) 2013

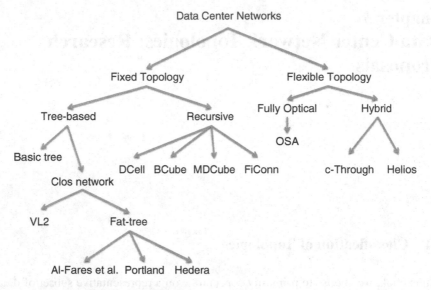

Fig. 3.1 A taxonomy of data center topologies

It should be noted that n and k may vary according to the position of the node. Servers and switches are regarded as vertices, wires as edges, allowing for the network topology to be represented as a graph.

3.2 Fixed Tree-Based Topologies I

Standard tree-based topologies and their variants are widely used for data center networks [2, 3, 18]. In this section, we review the basic tree first. Cisco's recommended tree-based topology discussed in Sect. 2.1.1 (See Fig. 2.2), is a $1 + 1$ redundant tree. Similarly Clos networks, which were reviewed in Sect. 2.1.2, is classified under tree-based topologies in this book. Indeed, a folded-Clos [6] can be viewed as a multi-rooted tree.

3.2.1 Basic Tree

Basic tree topologies consist of either two or three levels of switches/routers, with the servers as leaves. In a 3-level topology, there is a *core* tier at the root of the tree, an *aggregation* tier in the middle, and an *edge* tier of switches connecting to the servers. In a 2-level topology, there is no *aggregation* tier. There are no links between switches in the same tier, or in nonadjacent tiers. Figure 3.2 shows a 3-level tree topology.

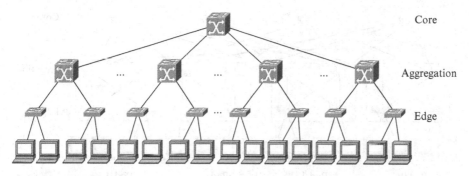

Fig. 3.2 A 3-level tree topology

In a basic tree topology, the higher-tier switches need to support data communication among a large number of servers. Thus switches with higher performance and reliability are required in these tiers. The number of servers in a tree topology is limited by the numbers of ports on the switches.

3.2.2   Fat-Tree

Fat-trees have been used as a topology for data centers by several researchers [2, 3, 18]. Here we need to make a distinction between the definition of a fat-tree as given by Leiserson [14] and the use of the term fat-tree for data center networks. A fat-tree as defined by Leiserson [14] is a network based on a complete binary tree, where the links connecting nodes in adjacent tiers become *fatter* as we go up the tree towards the root. This is achieved by constructing the topology such that for any switch, the number of links going down to its children is equal to the number of links going up to its parent, assuming each link is of the same capacity.

A DCN topology inspired by the fat-tree was first presented by Al-Fares et al. [2]. They use standard commodity Ethernet switches to implement the topology. In their approach, each n-port switch in the edge tier is connected to $\frac{n}{2}$ servers. The remaining $\frac{n}{2}$ ports are connected to $\frac{n}{2}$ switches in the aggregation level. The $\frac{n}{2}$ aggregation-level switches, the $\frac{n}{2}$ edge-level switches and the servers connected to the edge switches form a basic cell of a fat-tree, which is called a *pod*. In the core level, there are $\left(\frac{n}{2}\right)^2$ n-port switches, each one connecting to each of the n pods. Figure 3.3 shows a fat-tree topology with $n = 4$. Unlike in basic tree topology, all the three levels use the same kind of switches. High-performance switches are not necessary in the aggregation and core levels. The maximum number of servers in a fat-tree with n-port switches is $\frac{n^3}{4}$ [2].

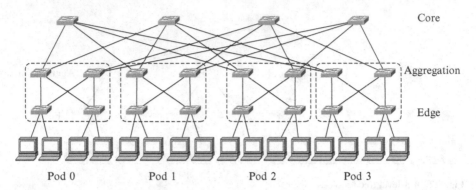

Fig. 3.3 A 3-level fat-tree topology

The topology thus realized is a special case of a folded-Clos network as acknowledged by Al-Fares et al. [2]. The term *fat-tree* however got associated with this topology and thereafter used in subsequent papers by the same research group [3, 18]. In order to avoid further confusion we refer to these topologies as fat-trees in this book. To emphasise their relationship to folded-Clos networks, we classify them as a subset of Clos networks in our taxonomy (see Fig. 3.1).

3.3 Fixed Recursive Topologies II

Tree-based topologies can be scaled up by inserting more levels of switches, while each server is connected to only one of the bottom level switches. The recursive topologies still have levels/tiers as in the tree-based topologies. However, recursive topologies use lower level structures as *cells* to build higher level structures, and the servers in recursive topologies may be connected to switches of different levels or even other servers. There are multiple network ports on the servers of recursive topologies, making them significantly different from the tree-based topologies. Graphs, rather than multi-rooted trees, are suitable to depict recursive topologies. Some representative recursive topologies are discussed in the following sections.

3.3.1 DCell

The most basic element of a DCell [12], which is called $DCell_0$, consists of n servers and one n-port switch. Each server in a $DCell_0$ is connected to the switch in the same $DCell_0$.

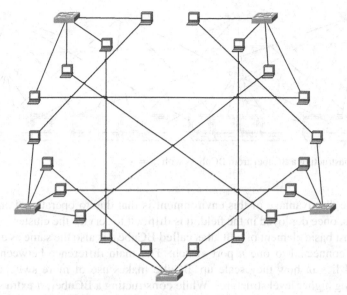

Fig. 3.4 Constructing a DCell$_1$ from DCell$_0$s with $n = 4$

Let DCell$_k$ be a level-k DCell. The first step is to construct a DCell$_1$ from several DCell$_0$s. Each DCell$_1$ has $n + 1$ DCell$_0$s, and each server of every DCell$_0$ in a DCell$_1$ is connected to a server in another DCell$_0$, respectively. As a result, the DCell$_0$s are connected to each other, with exactly one link between every pair of DCell$_0$s. A similar procedure is used to construct a DCell$_k$ from several DCell$_{k-1}$s. In a DCell$_k$, each server will eventually have $k + 1$ links: the first link or the level-0 link connected to a switch when forming a DCell$_0$, and level-i link connected to a server in the same DCell$_i$ but a different DCell$_{i-1}$. Assume that each DCell$_{k-1}$ has t_{k-1} servers, then a DCell$_k$ will consist of t_k DCell$_{k-1}$s, and consequently $t_{k-1} \times t_k$ servers. Obviously, we have $t_k = t_{k-1} \times (t_{k-1} + 1)$. Figure 3.4 shows a DCell$_1$ when $n = 4$. It can be seen that the number of servers in a DCell grows double-exponentially, and the total number of levels in a DCell is limited by the number of NICs on the servers in it. DCell can scale to very large number of servers using switches and servers with very few ports. For example, when $n = 6, k = 3$, a fully constructed DCell can comprise more than three million servers [12].

3.3.2 BCube

BCube [13] is a recursive topology specially designed for shipping container based modular data centers. Building a data center cluster in a 20- or 40-foot shipping container makes it highly portable. As demands change at different data centers, the whole cluster can be readily moved. While the deployment time is considerably

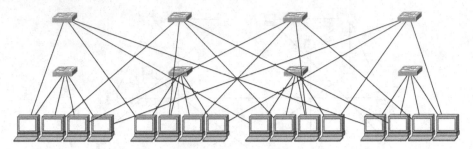

Fig. 3.5 Constructing a BCube$_1$ from BCube$_0$s with $n = 4$

shorter, the disadvantage of this environment is that due to operational and space constraints, once deployed in the field, it is difficult to service the cluster.

The most basic element of a BCube, called BCube$_0$, is also the same as a DCell$_0$: n servers connected to one n-port switch. The main difference between BCube and DCell lies in how they scale up. BCube makes use of more switches when constructing higher level structures. While constructing a BCube$_1$, n extra switches are used, connecting to exactly one server in each BCube$_0$. Therefore a BCube$_1$ contains n BCube$_0$s and n extra switches (If the switches in the BCube$_0$s are taken into consideration, there are totally $2n$ switches in a BCube$_1$). More generally, a BCube$_k$ is constructed from n BCube$_{k-1}$s and n^k extra n-port switches. These extra switches are connected to exactly one server in each BCube$_{k-1}$. In a level-k BCube, each level requires n^k switches (each of which is an n-port switch).

BCube makes use of more switches when constructing higher level structures, while DCell uses only level-0 n-port switches. Both however require servers to have $k + 1$ NICs. The implication is that servers will be involved in switching more packets in DCell than in BCube. Figure 3.5 shows a BCube$_1$ with $n = 4$.

Just like the DCell, the number of levels in a BCube depends on the number of ports on the servers. The number of servers in BCube grows exponentially with the levels, much slower than DCell. For example, when $n = 6$, $k = 3$, a fully constructed BCube can contain 1,296 servers. Considering that BCube is designed for container based data centers, such scalability is sufficient [13].

3.4 Flexible Topologies

Recently, researchers have proposed using optical switching technology to construct DCNs [5, 9, 20]. Besides offering high bandwidth (up to Tbps per fiber with WDM techniques), optical networks offer significant flexibility since the topology can be reconfigured during operation. Such a feature is important given the unbalanced and ever-changing traffic patterns in DCNs. In this section, we introduce three flexible-topology DCN architectures: c-Through [20], Helios [9] and OSA [5].

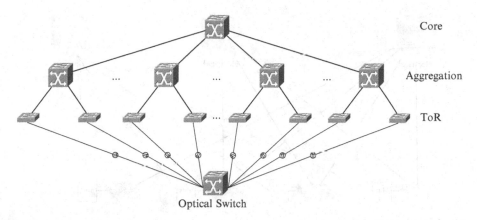

Fig. 3.6 c-Through network

3.4.1 c-Through

c-Through [20] is a hybrid network architecture that consists of both an electrical packet switching network and an optical circuit switching network. The hybrid network of c-Through, or the so-called HyPaC (Hybrid Packet and Circuit) Network, consists of two parts: a tree-based electrical network which maintains connectivity between each pair of ToR switches, and a reconfigurable optical network that offers high-bandwidth interconnection between pairs of racks as needed.

Due to the relatively high cost optical network and the high bandwidth of optical links, it is unnecessary and not cost-effective to maintain an optical link between each pair of racks; instead c-Through connects each rack to exactly one other rack at a time. As a result, the high-capacity optical links are offered to pairs of racks on a transient basis according to traffic demand. The estimation of traffic between racks and reconfiguration of the optical network is accomplished by the control plane of the system. Figure 3.6 shows a c-Through network (servers that are directly connected to ToR switches are not shown).

To configure the optical part of a c-Through network, the traffic between racks should be known a priori. c-Through estimates rack-to-rack traffic demands by observing the occupancy levels of socket buffers. Since a rack can have an optical-network connection to only one other rack at a given instant, connections should be made between racks that are estimated to exchange the most amount of traffic. In [20], the problem is solved using the maximum-weight perfect matching algorithm (Edmonds' Algorithm) [8]. The topology of the optical network is configured accordingly.

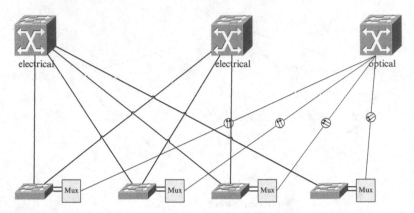

Fig. 3.7 Helios network

3.4.2 Helios

Helios [9] is another hybrid network with both electrical and optical switches. Helios is a 2-level multi-rooted tree of core and pod (ToR) switches. Core switches can be either electrical or optical so as to make full use of the two complementary techniques. Helios is similar to c-Through but it uses Wavelength Division Multiplexing (WDM). On each of the pod switches, the uplinks are equipped with a optical transceiver. Half of the uplinks are connected to the electrical switches, while the other half are connected to the optical switch through an optical multiplexer. Figure 3.7 shows a Helios network topology (the transceivers are not shown).

Unlike c-Through, Helios uses the electrical packet switching network to carry bursty traffic, while the optical circuit switching network offers baseline bandwidth to slow changing traffic. Helios uses an algorithm from Hedera [3] to estimate traffic demand between racks. The goal of the configuration of the optical network is the same as in c-Through: to maximize the traffic demand that can be satisfied. Helios also formulates the optimization problem as maximum-weight matching and solves it using Edmonds' Algorithm.

3.4.3 OSA

OSA [5] is a *pure* optical switching network, which means that it abandons the electrical core switches and uses only optical switches to construct the switching core. The ToR switches provide the interface between the server side electrical connections and the optical signals of the switching core. OSA allows multiple links to the switching core from each ToR switch (the typical number of such links is 4). However, the connection pattern is determined flexibly according to traffic demand. Since the network does not ensure a direct optical link between each pair of racks, the controlling system constructs a topology that makes it a connected graph in which a ToR switche becomes responsible for relaying traffic between two other ToR switches.

OSA estimates the traffic demand using the same method as Helios. However in OSA, there can be multiple optical links offered to each rack, and the problem can no longer be formulated as the maximum-weight matching problem. It is a multi-commodity flow problem with degree constraints, which is NP-hard. In [5], the problem is simplified to a maximum-weight b-matching problem [17], and solved approximately by applying multiple maximum-weight matchings.

3.5 Comparison of Topologies

3.5.1 Comparison of Scale

Table 3.1 presents a comparison of some of the parameters of the topologies introduced earlier. The flexible-topology architectures are not included. The following parameters are used when comparing the different topologies:

- *Degree of the servers*: The (average) number of network ports on the servers in the data center. For the tree-based topologies, only one port is needed on each server. However, in the recursive topologies, the degree of servers may vary according to the levels required. Considering the data shown in Table 3.2, k does not need to be large to scale up. On the other hand, since the servers are not specifically designed for switching, and the actual bandwidth that a server can offer is also limited by the speed of storage, it will be meaningless to add more and more ports on servers.
- *Diameter*: The longest of the shortest paths between two servers in a data center. A smaller diameter leads to more effective routing, and lower transmission latency in practice. For most topologies the diameter grows logarithmically with the number of servers.
- *Number of Switches*: It is assumed that all switches are the same in each topology. This assumption also stands for Tables 3.2, 3.3 and 3.5. BCube uses the largest number of switches, which may lead to higher cost. The basic tree topology uses the fewest switches because there is little hardware redundancy in the architecture.
- *Number of Wires*: This metric shows the number of wires required when deploying a data center. BCube uses the most wires, followed by DCell. However, it should be noted that the number of wires only shows the wiring complexity; it does not show the accurate wiring cost, because not all the topologies use homogeneous wires. E.g., the typical bandwidth of a link between an aggregation switch and a core switch in fat-tree is 10 Gbps, while it is 1 Gbps on a link between a server and an edge switch.

Table 3.1 Summary of parameters

	Tree-based architecture			Recursive architecture	
	Basic tree	Fat-tree [2]	Clos network [11]	DCell [12]	BCube [13]
Degree of servers	1	1	1	$k+1$	$k+1$
Diameter	$2\log_{\varepsilon_n-1} N$	6	6	$2^{k+1}-1$	$\log_n N$
No. of switches	n^2+n+1 N	$\dfrac{5N}{n}$	$\dfrac{3}{2}n+\dfrac{n^2}{4}+\dfrac{4N}{n_{ToR}}$	$\dfrac{N}{n}$	$\dfrac{N}{n}\log_n N$
No. of wires	$\dfrac{n}{n-1}(N-1)$	$N\log_{\frac{n}{2}}\dfrac{N}{2}$	$N+\dfrac{n^2}{4}\times n_{ToR}$	$\left(\dfrac{k}{2}+1\right)N$	$N\log_n N$
No. of servers	$(n-1)^3$	$\dfrac{n^3}{4}$	$\dfrac{n^2}{4}\times n_{ToR}$	$\geq (n+\frac{1}{2})^{2^k}-\frac{1}{2}$, $\leq (n+1)^{2^k}-1$	n^{k+1}

Table 3.2 Number of servers

	Tree-based topology				Recursive topology	
n	Basic tree	Fat-tree [2]	Clos network [11]	k	DCell [12]	BCube [13]
				2	420	64
4	64	16	8	3	176, 820	256
				4	$>3 \times 10^{10}$	1, 024
				2	1, 806	216
6	216	54	36	3	3, 263, 442	1, 296
				4	$>10^{13}$	7, 776
				2	5, 256	512
8	512	128	96	3	27, 630, 792	4, 096
				4	$>7 \times 10^{14}$	32, 768
				2	74, 256	4, 096
16	4, 096	1, 024	896	3	$>5 \times 10^{9}$	65, 536
48	110, 592	27, 648	26, 496	2	5, 534, 256	110, 592

Table 3.3 Performance summary

	Tree-based topology			Recursive topology	
	Basic tree	Fat-tree [2]	Clos network [11]	DCell [12]	BCube [13]
One-to-one	1	1	1	$k+1$	$k+1$
One-to-several	1	1	1	$k+1$	$k+1$
One-to-all	1	1	1	$k+1$	$k+1$
All-to-all	n	N	$\dfrac{2N}{n_{ToR}}$	$>\dfrac{N}{2^k}$	$\dfrac{n}{n-1}(N-1)$
Bisection width	$\dfrac{n}{2}$	$\dfrac{N}{2}$	$\dfrac{N}{n_{ToR}}$	$>\dfrac{N}{4\log_n N}$	$\dfrac{N}{2}$

- *Number of Servers*: All the metrics above are computed under the same number of servers (N), while this one shows the scalability of different topologies with the same n and k (for recursive topologies only). It is assumed in this row that the tree-based topologies use 3-level structure. Considering the data in Table 3.2, it is no doubt that DCell scales up much faster than other architectures for the number of servers in DCell grows double-exponentially with k.

3.5.2 Comparison of Performance

Table 3.3 shows the comparison of some performance metrics of different topologies.

- *Bandwidth*: The first four rows of Table 3.3 show the bandwidth that the topologies can offer under different traffic patterns. "One-to-one" means the maximum bandwidth that the topology can offer when one arbitrary server sends data to another arbitrary server, and so on so forth. "All-to-all" bandwidth means every server establishes a flow to all the other servers. Each kind of these traffic pattern is meaningful for different applications. E.g., one-to-several traffic occurs when the file system is making replicas, one-to-all traffic occurs when updating some software on all the servers, and all-to-all traffic is very common in MapReduce.

Here in Table 3.3 the bandwidths are expressed as the number of links, which implies that each link in a data center has the same bandwidth. It is shown in the table that one-to-one, one-to-several and one-to-all bandwidths are in fact limited by the number of ports on the servers, or the degree of servers. The basic tree topology offers the smallest all-to-all bandwidth, limited by the number of ports of the switch at the root. Since fat-tree introduces redundant switches, it can offer as many edge-disjoint paths as the number of servers.

- *Bisection Width*: The minimum number of wires that must be removed when dividing the network into two equal sets of nodes. The larger the bisection width, the better fault-tolerance ability of a topology. A similar metric, the bisection bandwidth is also widely used, which means the minimum collective bandwidth of all the wires that must be removed to divide the network into two equal sets of nodes. Basic tree has the smallest bisection width, which means that it is the most vulnerable of all.

A graceful degradation of performance implies that when more and more components fail in the data center, performance reduces slowly without a steep decline in performance. When building a new data center, it is usual that a partial topology will be built first, and more components are added based on future need. This partial network can be viewed as a network with many failures. Building a partial network with a topology possessing the property of graceful degradation means that people can get expected performance with fewer components. Guo et al. [13] give a comparison between fat-tree, DCell and BCube on this metric with simulation results.

3.5.3   Performance Evaluation Using Simulation

We compared some of the topologies discussed above using a simulation tool DCNSim [16]. Using simulation it is possible to compute some metrics which are not as explicit as the metrics shown in Sects. 3.5.1 and 3.5.2. Such metrics include:

- *Aggregated Bottleneck Throughput (ABT)*. Cloud applications such as VM migration [18] and distributed file system [10] requires high throughput during operation. To evaluate throughput of the whole network, we use *ABT* [13] to evaluate the maximum sustainable throughput over the entire DCN. Given multiple flows over a link, the bandwidth of this link is shared by these flows uniformly, and the maximum throughput between any pair of servers is restricted to the bottleneck throughput of the corresponding route. The formula for calculating the *ABT* is:

Table 3.4 Simulation results

	Fat-tree [2]	DCell [12]	BCube 3 [13]	BCube 5
Degree of switches	24	7	15	5
Degree of servers	1	3	3	5
No. of servers	3,456	3,192	3,375	3,125
Aggregated bottleneck throughput	3,456	1,003.3	3,615	3,905
Average path length	5.911	9.092	5.602	8.003

$$ABT = \sum_{i,j} \left(\min_{L \in P(s_i, s_j)} \frac{B_L}{F_L} \right)$$

Here $P(s_i, s_j)$ denotes the path between two servers s_i and s_j when s_i is sending a flow to s_j, while B_L and F_L denotes the bandwidth and the total number of flows running concurrently on link L.

- *Average Path Length (APL)*. In order to measure the latency of traffic, we use hop-count based path length to denote the transmission delay of a path. For the overall performance of the whole network, *APL* is used for evaluation.

Some of the results are shown in Table 3.4. The selected topologies are Fat-Tree, DCell and BCube. To be comparable, the topologies are configured with similar number of servers. As a special case, BCube has two different configurations with similar number of servers. Both metrics are evaluated under all-to-all traffic, i.e., each server communicates with every other server, which is the heaviest traffic that can be generated in a network. Totally there will be $N \times (N - 1)$ flows. The routing algorithm we use is shortest path routing.

3.5.4 Hardware Redundancy of Data Center Network Topologies

The hardware redundancy discussed here is based on the topology of data center network. Some data center network architectures, such as PortLand and VL2, which requires more reliability for their centralized management system, are not considered here.

Table 3.5 shows the hardware redundancy offered by different topologies. Before discussing the metrics presented in the table, the definitions of some of the terms are given below:

Definition 3.1 (Node-disjoint Paths). The minimum of the total number of paths that share no common intermediate nodes between any arbitrary servers.

Definition 3.2 (Edge-disjoint Paths). The minimum of the total number of paths that share no common edges between any arbitrary servers.

Table 3.5 Hardware redundancy summary

			Tree-based topology			Recursive topology	
			Basic tree	Fat-tree	Clos	DCell	BCube
Node-disjoint paths			1	1	1	$k+1$	$k+1$
Edge-disjoint paths			1	1	1	$k+1$	$k+1$
Redundancy	Switch	Edge/ToR	0	0	0	k	k
		Aggregation	0	$\frac{n}{2}-1$	1		
		Core/Intermediate	0	$\frac{n^2}{4}-1$	$\frac{n}{2}-1$		
	Server		-	-	-	k	k
Level							

Definition 3.3 (f-fault-tolerance [7]). A network is f-fault-tolerant if for any f failed components in the network, the network is still connected.

Definition 3.4 (Redundancy Level [7]). A network has a redundancy level equal to r if and only if, after removing any set of r links, it remains connected, and there exists a set of $r + 1$ links such that after removing them, the network is no longer connected.

In the table, the redundancy levels of different components are shown respectively, which reflects the influence of different failures.

The numbers of node-disjoint paths and edge-disjoint paths are the same in each of the topology. This is because there is exactly one link between each pair of directly connected nodes (server to server, server to switch, switch to switch) in these topologies. If redundant links are added making the topology a multigraph, there will be more edge-disjoint paths than node-disjoint paths. The number of node-disjoint paths shows the ability of a system to deal with node failures, while the number of edge-disjoint paths shows the ability of dealing with link failures. A node failure can have a more severe impact than a link failure, because it may lead to several link failures.

The tree-based topologies have fewer number of node-disjoint paths and edge-disjoint paths compared to DCell and BCube. The main reason for this is that tree-based topologies place fewer ports on the servers. Each server is connected to an edge switch with single connection in tree-based topologies. Fewer ports on servers makes the connections between servers and switches vulnerable to failures. For example, in the tree-based topologies, if one edge switch fails, every server connected to it will be separated from the remaining of the data center. This also explains why there is no redundancy of edge switches.

DCell and BCube, however, have more node-disjoint paths and edge-disjoint paths because of the multiple ports on their servers, which results in higher redundancy level. Unless each of the node-disjoint paths has a failure on it, two servers will always remain connected. It should be noted that the redundancy level of DCell or BCube is decided by the number of ports on the servers.

This table does not show the difference between the failure rate of different components. For example, in the basic tree topology, designers tend to use more expensive and reliable switches as the core switches, which have a much lower failure rate compared to the cheaper edge switches. We have more detailed results on fault-tolerance regarding software mechanisms in Sect. 6.2.7.

3.6 Potential New Topologies

Most of the topologies discussed above originated in the area of supercomputing or parallel computing. e.g., the fat-tree [14] was first proposed for supercomputing, and BCube is closely related to Hypercube [4], which was designed for parallel computing. This is because supercomputing and parallel computing share many design goals with data center networks, such as the need to interconnect large numbers of computing nodes, offer high-bandwidth connectivity, and support large volumes of data transfer.

The taxonomy that we propose is not the only approach for classifying data center network topologies. Zhang et al. [22] proposes a taxonomy by classifying data center networks into two categories, *hierarchical* topologies and *flat* topologies, which correspond to the tree-based topologies and recursive topologies in our survey respectively. Lucian Popa et al. classify topologies into *switch-only* and *server-only* topologies [19].

3.7 Summary

In this chapter, we reviewed several of the data center network topologies that have been proposed in the literature. These new topologies were designed to address some of the shortcomings of the current state-of-the-art DCN. We classified the topologies into fixed and flexible topologies, based on the ability to reconfigure the topology at run-time. We reviewed several salient features of each of the topologies. We then presented a comparison of the topologies based on their scaling properties, performance and hardware redundancy characteristics.

References

1. Abu-Libdeh, H., Costa, P., Rowstron, A., O'Shea, G., Donnelly, A.: Symbiotic routing in future data centers. ACM SIGCOMM Comput. Commun. Rev. **40**(4), 51–62 (2010)
2. Al-Fares, M., Loukissas, A., Vahdat, A.: A scalable, commodity data center network architecture. In: Proceedings of the ACM SIGCOMM 2008 Conference on Data Communication, Seattle, pp. 63–74. ACM (2008)
3. Al-Fares, M., Radhakrishnan, S., Raghavan, B., Huang, N., Vahdat, A.: Hedera: dynamic flow scheduling for data center networks. In: Proceedings of the 7th USENIX Conference on Networked Systems Design and Implementation, San Jose, p. 19. USENIX Association (2010)
4. Bhuyan, L., Agrawal, D.: Generalized hypercube and hyperbus structures for a computer network. IEEE Trans. Comput. **C-33**(4), 323–333 (1984)
5. Chen, K., Singla, A., Singh, A., Ramachandran, K., Xu, L., Zhang, Y., Wen, X., Chen, Y.: OSA: an optical switching architecture for data center networks with unprecedented flexibility. In: Proceedings of the 9th USENIX Conference on Networked Systems Design and Implementation, Berkeley, pp. 1–14. USENIX Association (2012)
6. Dally, W., Towles, B.: Principles and Practices of Interconnection Networks. Morgan Kaufmann, Amsterdam/San Francisco (2004)
7. Duato, J., Yalamanchili, S., Ni, L.: Interconnection Networks: An Engineering Approach. Morgan Kaufmann, San Francisco (2003)
8. Edmonds, J.: Paths, trees, and flowers. Can. J. math. **17**(3), 449–467 (1965)
9. Farrington, N., Porter, G., Radhakrishnan, S., Bazzaz, H., Subramanya, V., Fainman, Y., Papen, G., Vahdat, A.: Helios: a hybrid electrical/optical switch architecture for modular data centers. In: ACM SIGCOMM Computer Communication Review, New Delhi, vol. 40, pp. 339–350. ACM (2010)

10. Ghemawat, S., Gobioff, H., Leung, S.T.: The Google file system. SIGOPS Oper. Syst. Rev. **37**(5), 29–43 (2003). DOI http://doi.acm.org/10.1145/1165389.945450
11. Greenberg, A., Hamilton, J.R., Jain, N., Kandula, S., Kim, C., Lahiri, P., Maltz, D.A., Patel, P., Sengupta, S.: VL2: a scalable and flexible data center network. SIGCOMM Comput. Commun. Rev. **39**(4), 51–62 (2009). DOI http://doi.acm.org/10.1145/1594977.1592576
12. Guo, C., Wu, H., Tan, K., Shi, L., Zhang, Y., Lu, S.: DCell: a scalable and fault-tolerant network structure for data centers. ACM SIGCOMM Comput. Commun. Rev. **38**(4), 75–86 (2008)
13. Guo, C., Lu, G., Li, D., Wu, H., Zhang, X., Shi, Y., Tian, C., Zhang, Y., Lu, S.: BCube: a high performance, server-centric network architecture for modular data centers. ACM SIGCOMM Comput. Commun. Rev. **39**(4), 63–74 (2009)
14. Leiserson, C.E.: Fat-trees: universal networks for hardware-efficient supercomputing. IEEE Trans. Comput. **34**(10), 892–901 (1985)
15. Li, D., Guo, C., Wu, H., Tan, K., Zhang, Y., Lu, S.: Ficonn: using backup port for server interconnection in data centers. In: IEEE INFOCOM, Rio de Janeiro (2009)
16. Liu, Y., Muppala, J.: DCNSim: a data center network simulator. In: 33rd International Conference on Distributed Computing Systems Workshops (ICDCSW) 2013, Philadelphia, pp. 1–6. IEEE (2013)
17. Müller-Hannemann, M., Schwartz, A.: Implementing weighted b-matching algorithms: insights from a computational study. J. Exp. Algorithmics (JEA) **5**, 8 (2000)
18. Niranjan Mysore, R., Pamboris, A., Farrington, N., Huang, N., Miri, P., Radhakrishnan, S., Subramanya, V., Vahdat, A.: Portland: a scalable fault-tolerant layer 2 data center network fabric. ACM SIGCOMM Comput. Commun. Rev. **39**(4), 39–50 (2009)
19. Popa, L., Ratnasamy, S., Iannaccone, G., Krishnamurthy, A., Stoica, I.: A cost comparison of datacenter network architectures. In: Proceedings of the 6th International Conference, Philadelphia, p. 16. ACM (2010)
20. Wang, G., Andersen, D., Kaminsky, M., Papagiannaki, K., Ng, T., Kozuch, M., Ryan, M.: c-Through: part-time optics in data centers. In: ACM SIGCOMM Computer Communication Review, New Delhi, vol. 40, pp. 327–338. ACM (2010)
21. Wu, H., Lu, G., Li, D., Guo, C., Zhang, Y.: MDCube: a high performance network structure for modular data center interconnection. In: Proceedings of the 5th International Conference on Emerging Networking Experiments and Technologies, Rome, pp. 25–36. ACM (2009)
22. Zhang, Y., Sub, A., Jiang, G.: Understanding data center network architectures in virtualized environments: a view from multi-tier applications. Comput. Netw. **55**, 2196–2208 (2011)

Chapter 4
Routing Techniques

4.1 Introduction

Given a network topology, routing techniques and protocols determine how data packets are delivered from a source to a destination within a data center. The *orchestration* of the network in order to achieve the goals of data delivery among the servers is what we discuss in this chapter. *Addresses* are used to uniquely identify interfaces (or nodes) within a data center network. *Forwarding* is the mechanism by which packets are forwarded from one link to another at switches/routers. Typically this action is based on the destination address carried in a packet header. The *data-plane protocol* used for packet forwarding specifies the addressing format used. Also of interest is the *node at which forwarding actions* take place. In most cases, these actions occur at switches/routers, but in several of the architectures compared here, servers perform packet forwarding between the several NICs with which they are equipped. Such forwarding in servers can be implemented either in software or in specialized hardware. Finally, the *routing mechanisms* used to determine the next-hop node, whether executed on a packet-by-packet basis, or used in conjunction with routing protocols for pre-configuration of forwarding tables, are presented for each architecture. The focus of this section is on unicast, single-path routing, while multi-path routing will be discussed in Chap. 5. Now we present the details of the addressing, forwarding and routing mechanisms used in the different network topologies described in Chap. 3.

4.2   Fixed Tree-Based Topologies I

4.2.1   Fat-Tree Architecture of Al-Fares et al. [1]

4.2.1.1   Addressing

Al-Fares et al. proposed an architecture [1] that uses the dotted decimal notation of IP addresses to specify the position of a node (either a server or a switch) in the tree. As described in Sect. 3.2.2, a fat-tree has n pods, with n switches in each pod, while $\frac{n}{2}$ servers are connected to each of the $\frac{n}{2}$ edge switches in a pod. The reserved private 10.0.0.0/8 block is allocated to the nodes. The pod switches, including edge switches and aggregation switches are addressed in the form 10.*pod*.*switch*.1, where *pod* $\in [0, n-1]$ denotes the number of the pod that a switch belongs to, and *switch* $\in [0, n-1]$ denotes the position of a switch in a pod. The $\frac{n^2}{4}$ core switches are addressed as 10.*k*.*j*.*i*, where $i, j \in \left[1, \frac{n}{2}\right]$.

The $\frac{n}{2}$ servers connected to edge switch 10.*pod*.*switch*.1 is addressed as 10.*pod*.*switch*.*ID*, where $ID \in \left[2, \frac{n}{2} + 1\right]$ denotes the position of the server in the subnet connected to the switch. With such an addressing scheme, n can be as large as 256, supporting as many as 4.19M servers.

4.2.1.2   Forwarding and Routing

In the fat-tree architecture shown in Sect. 3.3, the core switches need to transmit all the inter-pod traffic. If hop count is used as the metric of evaluating a route, there will be $\left(\frac{n}{2}\right)^2$ shortest routes between two servers in different pods, each route with six hops: three hops from the source up to a core switch, and three hops from the core switch down to the destination. The main goal of routing in fat-tree is to distribute the traffic evenly among these core switches.

Al-Fares et al. [1] propose a 2-level routing table to solve this problem. Unlike the common routing table in which one entry relates several IP address with one port on the switch, every entry in this 2-level routing table may potentially have a pointer to a small secondary table. The entries of the secondary table are in the form of (*suffix*, *port*), while the entries in the primary table have the form (*prefix*, *port*). Not all the entries of the primary table have a pointer to a secondary entry, while one secondary entry may be pointed to by multiple primary entries. A primary entry is *terminating* if it points to no secondary entry. When routing a packet, the destination IP address will be first checked to match the longest prefix in the primary table. If the resulting primary entry points to a secondary table, the destination IP will be checked again to match the longest suffix, and the output port will be finally decided.

In such a two-level routing scheme, the pod switches have terminating prefixes for the servers in that pod, and a default /0 prefix with a secondary table for addresses outside the pod. As a result, in a pod switch, the entries in the primary table will distinguish destination servers that do not belong to the same pod, and the entries in a secondary table directs packets to different paths so as to get a balanced flow distribution, e.g., the aggregation switches will transfer packets with different suffixes through different core switches so that each core switch will receive the same number of packets in all-to-all traffic. For the core switches, the routing tables are simpler. They are assigned terminating prefixes for all network addresses, so that they can correctly direct every packet to the pod in which the destination server is located. Because one core switch has exactly one link to an aggregation switch in each pod, and one aggregation switch has exactly one link to every edge switch in its pod, when a packet arrives at a core switch, the remainder of its path to the destination is unique. Traffic diffusion happens only on the first half of a path, i.e., before a packet reaches a core switch.

It should be noted that in such a routing scheme, the original routing table is generated according to the architecture, not by any autonomous route discovering mechanism.

At first glance, it appears that this architecture proposes a direct use of IP routers in a fat-tree topology, and specifies a particular address assignment scheme that leverages the regular topology eliminating the need for a routing protocol such as OSPF. In other words, a simple algorithm can be executed to determine the output port based on the destination IP address given that the algorithm leverages the structured manner in which addresses are assigned.

However, on closer examination, one sees that the nodes at which packet forwarding occurs cannot be off-the-shelf IP routers performing standard longest-prefix matching operations. Instead a new router design[1] is required for a two-level lookup. Therefore, even though the data-plane protocol is unchanged, i.e., servers generate IP datagrams carried in Ethernet frames, new routers need to be designed, implemented, and deployed to realize this architecture.

As for the routing mechanism, a centralized controller is envisioned to create the two-level forwarding tables at all the routers in the fat-tree. Packet forwarding action is reduced to a two-level table lookup. No algorithm needs to be executed on a packet-by-packet basis to determine output port for packet forwarding as is required in some of the other architectures. The need for a distributed routing algorithm to handle failures is mentioned without details.

---

[1] Al-Fares et al. [1] refer to the nodes as *switches* even though packet forwarding action is at the IP layer; the term *router* is more appropriate.

4.2.2 PortLand and Hedera: Layer-2 Network Based on a Tree Topology

PortLand [6] offers layer-2 addressing, forwarding and routing in multi-rooted tree based data centers. Hedera [2] is based on PortLand, and its implementation augments PortLand routing and fault-tolerant protocols. PortLand/Hedera use a centralized solution to manage the whole fabric.

4.2.2.1 Addressing in PortLand

PortLand uses a 48-bit hierarchical Pseudo MAC (PMAC) addressing scheme. A unique PMAC is assigned to each machine (physical or virtual), encoding the location of it. For example, in the fat-tree architecture, all the servers inside the same pod will have the same prefix in the PMAC addresses. Each server still keeps its actual MAC (AMAC) address. The ARP requests received by the fabric manager will return the PMAC address corresponding to an IP address. The destination MAC address in the Ethernet frame will correspond to the PMAC address until it reaches the edge switch, where the mapping between the destination's PMAC and AMAC addresses is stored.

4.2.2.2 Fabric Manager

The fabric manager of PortLand is a user process running on a dedicated computer inside the data center. It is responsible for assisting in ARP resolutions, and supporting fault-tolerance, multicast and other functions. This centralized solution can help reduce the complexity of routing, but it simultaneously decreases robustness. To keep robustness high, the amount of centralized information stored at the fabric manager is limited, and administrator configuration of the manager is not required. Besides, one or more replicas of the fabric manager are stored with asynchronous updates.

4.2.2.3 Proxy-Based ARP

Address Resolution Protocol (ARP) is used to map IP addresses to MAC addresses. However, in PortLand, since PMAC is used, standard ARP does not work. In PortLand, the ARP requests sent by a server are intercepted by the edge switch connecting it, and the edge switch will forward the requests to the fabric manager. The fabric manager will look up in its PMAC-IP table, and if there is a matched entry, it will return the result to the edge switch, which will finally create an ARP reply to the requesting server.

It is possible that the fabric manager does not have a matched entry for a request, for example, in the case of failed recovery. When this happens, the fabric manager will have to launch a broadcast to retrieve the mapping. The host owning the target IP will reply with its AMAC, which will be translated to its PMAC by the connecting switch.

PortLand offers an additional feature for more convenient VM migration. While the AMAC will remain the same no matter how a VM migrates, its PMAC and IP address will change according to its physical position. After migration is completed, a VM will send an ARP message to the fabric manager reporting its new IP address. The fabric manager will then send an ARP invalidation message to the VM's previous switch. For those machines that are still attempting to communicate with the previous PMAC of the migrated VM, this previous switch will send a unicast ARP message to each of these machines to announce the new PMAC and IP of the migrated VM.

4.2.2.4 Location Discovery and Packet Forwarding

PortLand uses a Location Discovery Protocol (LDP) to discover the locations of the switches, so as to make packet forwarding and routing more effective. A PortLand switch will only forward data packets after its location is established. It will periodically send a Location Discovery Message (LDM) through all its ports. A LDM contains the ID of the switch and its location information. LDMs help switches to find their neighboring switches, and locate themselves by checking the number and states of switches to which they are connected. Finally the fabric manager will assign a number corresponding to the location of each of the switches.

Once the location of a switch is established by means of LDP, it can use the location information to forward packets. For example, in the fat-tree architecture, a core switch can learn pod numbers from the aggregation switches to which it is connected. When a packet arrives, it needs to inspect the bits corresponding to pod number in the PMAC of the packet, and decides on which port to forward the packet.

LDP offers an automated mechanism for switches to discover their own locations, and thus eliminates the need for manual configuration. This is meaningful for the initial construction of a data center, and for recovery from failures.

4.2.3 VL2

VL2 [3] is based on a folded-Clos network topology. This architecture inserts a layer-2.5 shim into the servers' network protocol stack to implement features unavailable in network adapters/drivers. Addressing is IP based. The protocols are unchanged, i.e., Ethernet and IP are used. Switches and routers are used unchanged. The change required is the addition of a new layer between IP and MAC in end host's networking software stacks.

4.2.3.1   Addressing in VL2

Similar to PortLand, VL2 is designed to maintain flexibility of IP address assignment to services or VMs so that servers can be reassigned to services according to requirements, without changing the actual network architecture. When a server needs to send a packet, the shim layer inserted into its protocol stack will inquire a directory system for the actual location of the destination, then rewrite the address in the packet header and tunnel the packet to the shim layer of the destination.

To achieve the design goal above, VL2 makes use of two IP address families: the location-specific IP addresses (LAs) for network infrastructure, and the application-specific IP addresses (AAs) for applications. LAs are assigned to the switches and interfaces on servers, and are used to forward packets. The AAs, however, are assigned to applications, and remain unchanged even if the physical location of the server changes or VMs migrate. Each AA is related to an LA. The VL2 directory system is implemented to store the mapping from AAs to LAs and answer queries. The shim layer on each servers, or so-called VL2 agent, is responsible for replacing AAs with LAs when sending packets, and LAs back to AAs when receiving packets.

4.2.3.2   VL2 Directory System

VL2 uses a centralized management system called the Directory System to offer necessary address mapping services required by the infrastructure. The system consists of two types of servers:

- A modest number (50–100 servers for 100 K servers) of replicated *directory servers*, which cache address mappings and reply to queries;
- A small number (5–10 servers) of *replicated state machines (RSMs)*, which offer a reliable storage for address mappings.

Each of the directory servers keeps a copy of the whole AA-LA mappings, and uses its local information to reply to queries from servers. To maintain consistency, a directory server periodically synchronizes its mappings with an RSM. When there is an update in the mapping, the update will be sent to a random directory server, which will forward the update to an RSM server. The RSM server will replicate the update to each of the RSMs, and reply to the directory server, which then forwards the reply to the originating machine.

4.2.3.3   Packet Forwarding in VL2

Since the traffic between servers use AAs, while the underlying network only knows routes for LAs, the VL2 agent at each server has to trap each packet and encapsulate the packet with an outer header that carries the LA address of the ToR switch of the destination. When the packet arrives at the TOR switch, it will be decapsulated and sent to the destination AA.

4.3   Fixed Recursive Topologies II

4.3.1   DCell

4.3.1.1   Addressing in DCell

Every server in a $DCell_k$ is identified with a $(k + 1)$-tuple $[a_k, a_{k-1}, \cdots, a_1, a_0]$, where a_k indicates which $DCell_k$ the server belongs to, and a_0 is the number of the server in its $DCell_0$. With a tuple, a server can be identified by a unique ID $uid_k \in [0, t_k)$ (t_k is the number of servers in a $DCell_k$), by simply calculating $uid_k = a_0 + \sum_{j=1}^{k} (a_j \times t_{j-1})$.

With its ID, a server in $DCell_k$ can be given a recursive defined expression as $[a_k, uid_{k-1}]$, where a_k is the number of the $DCell_{k-1}$ that the server belongs to, and uid_{k-1} is its ID in its $DCell_{k-1}$. In [4], DCell defines its own packet header in which 32-bit addressing is used.

4.3.1.2   DCellRouting

DCellRouting [4] is a simple routing algorithm for unicast, making use of the structure of DCell, and the addressing scheme discussed above. However, it does not take potential failures into consideration, which will be addressed in an updated version of DCellRouting. DCellRouting uses a divide-and-conquer approach. Assume that the source and the destination are in the same $DCell_k$, but in different $DCell_{k-1}$s. The algorithm will first find the link (n_1, n_2) connecting the two $DCell_{k-1}$s, and thus the routing is divided into two sub-paths: from source to n_1, and from n_2 to destination. The procedure is repeated until finally all the sub-paths are direct links. The maximum path length in DCellRouting is at most $2^{k+1} - 1$ [4]. Considering that a small k is sufficient to support a large number of nodes, this algorithm can find the route quite fast.

Although the average length of paths found by DCellRouting is quite close to the shortest path [4], DCellRouting is not a shortest-path routing scheme. This can be shown in Fig. 4.1. The shortest path between the source and the destination is of length 4, while the route decided by DCellRouting is of length 5. This happens when the source and destination, although not in the same $DCell_{k-1}$, are both connected to another $DCell_{k-1}$. In such cases the shortest path between two nodes in different $DCell_{k-1}$s is not via the link connecting the two $DCell_{k-1}$s they belong to, but via the $DCell_{k-1}$ to which they are both connected.

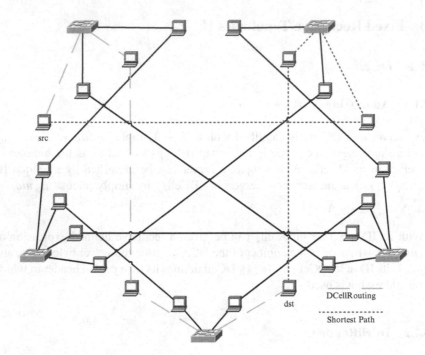

Fig. 4.1 DCellRouting and shortest path

4.3.1.3 DCellBroadcast

DCellBroadcast is a robust broadcasting scheme for DCell. In DCellBroadcast, when a server initializes a broadcast, it will send the packet to all its $k + 1$ neighbors. When receiving a broadcast packet, a server will check whether it has already received the same packet earlier; if not, it will broadcast the packet to all its neighbors. The steps for a broadcast packet to reach every server in the DCell is limited by the diameter (the longest of all the shortest paths between two nodes) of the DCell, $2^{k+1} - 1$. This scheme generates duplicate broadcast packets, but is robust since it can send the packet to every server as long as the network is connected.

4.3.2 BCube

4.3.2.1 Addressing in BCube

As mentioned in Sect. 3.3.2, each BCube$_l$ contains n BCube$_{l-1}$s, and each BCube$_0$ contains n servers. In [5], a server in a BCube is identified with an address array $a_k a_{k-1} \cdots a_0$, where $a_i \in [0, n-1]$, $i \in [0, 1)$. An equivalent way of determining this address array is calculating the *BCube Address* $baddr = \sum_{i=0}^{k} a_i n^i$. A switch is similarly identified in BCube: $< l, s_{k-1} s_{k-2} \cdots s_0 >$, where $s_j \in [0, n-1]$, $j \in [0, k-1]$, $0 \leq l \leq k$ is the level of the switch.

4.3.2.2 BCubeRouting: Single-Path Routing in BCube

The structure of BCube and the addressing scheme offer a simple way to implement single-path routing in BCube. In the addressing scheme, servers connected to the same switch differ in only one digit in their addresses, i.e., the Hamming distance between them is 1. Each packet includes a BCube header between the Ethernet header and IP header. When forwarding a packet from source to destination, the intermediate nodes will determine the digits in which their addresses differ, and change one digit in each step. This implies that the packet is transmitted from one server to another through the switch connecting them. These intermediate switches are uniquely determined as long as the order of changing the digits is determined. In this routing scheme, a packet travels through at most $k+1$ switches and k servers to arrive at the destination.

4.4 Summary

In this section, several routing mechanisms designed for different architectures are discussed. Different schemes are compared below, and from this comparison a few basic guidelines for designing path selection algorithms for a data center network are derived. Table 4.1 shows a comparison of some properties of different architectures.

4.4.1 Addressing

An addressing scheme that leverages the regular architecture of a data center network can lead to effective routing. It has to be scalable because of the large number of servers in a data center. It does not necessarily have to follow the addressing schemes used in the Internet at large.

Table 4.1 Routing summary

	Fat-tree [1]	DCell [4]	BCube [5]	Portland [6]	VL2 [3]
Address type	IP	MAC/IP	MAC/IP	MAC	MAC
Address resolution	Distributed	Distributed	Distributed	Centralized	Centralized
Route decision	Intermediate	Source	Source	Intermediate	Intermediate
Compatibility	Specialized	Specialized	Specialized	Compatible	Compatible

Most architectures discussed mainly use switches, eliminating routers. Eliminating routers makes it possible to apply layer-2 addressing and switching in data centers, as illustrated by Sect. 4.2.3. It should be noted that the addressing schemes discussed in Sects. 4.3.1 and 4.3.2 are not necessarily layer-3 addressing; they can be expressed in either MAC addresses or IP addresses, using routing protocols running at the corresponding layer. When routing with layer-3 addresses, the protocol stack does not need to be different than when forwarding on Ethernet headers as the IP headers can be carried transparently. However, switching with layer-2 addresses eliminates the need for intermediate nodes to analyze every packet to extract layer-3 addresses and re-encapsulate packets within layer-2 frames. In data centers, where high-speed and scalable routing is required, it is meaningful to design a new protocol stack if it actually enhances the performance of the network.

4.4.2 Centralized and Distributed Routing

Except PortLand and VL2, all the other routing schemes discussed use a distributed approach. On the other hand, Portland and VL2 can be applied to any multi-rooted tree architecture [3, 6]. The architecture itself does not restrict routing to a centralized or distributed approach. Each approach has its own advantages and disadvantages. Here some of the advantages of each approach are summarized.

Advantages of centralized routing include:

1. Optimized routing: The centralized system has the complete view of the entire data center, enabling it to make optimal routing decisions thus ensuring better use of the hardware.
2. Greater ability to deal with failures: A centralized system can gather failure information from the entire data center, so that it is easier to judge the effects of a failure and to enable mechanisms to handle it.
3. Easier configuration: Centralized routing protocols are deployed on one or several nodes inside a data center. Only these nodes need to be reconfigured when applying a different routing algorithm.

Advantages of distributed routing include:

1. Robustness: The routing function and information is distributed across all the servers in the data center. The routing protocol can operate as long as the data center remains connected.
2. More effective routing decision: Routing decisions are made according to local and neighbor node information only, saving the time and effort needed to gather information about the entire network.
3. Less overhead: Servers do not have to exchange routing information across a wide area so that bandwidth can be saved.

The two approaches are not mutually exclusive. For example, there is a distributed location discovery mechanism in the centralized PortLand [6]. Most distributed routing protocols will gather information from servers inside a certain area to make a "centralized" routing decision in the area.

References

1. Al-Fares, M., Loukissas, A., Vahdat, A.: A scalable, commodity data center network architecture. In: Proceedings of the ACM SIGCOMM 2008 Conference on Data Communication, Seattle, pp. 63–74. ACM (2008)
2. Al-Fares, M., Radhakrishnan, S., Raghavan, B., Huang, N., Vahdat, A.: Hedera: dynamic flow scheduling for data center networks. In: Proceedings of the 7th USENIX Conference on Networked Systems Design and Implementation, Berkeley, p. 19. USENIX Association (2010)
3. Greenberg, A., Hamilton, J.R., Jain, N., Kandula, S., Kim, C., Lahiri, P., Maltz, D.A., Patel, P., Sengupta, S.: VL2: a scalable and flexible data center network. SIGCOMM Comput. Commun. Rev. 39(4), 51–62 (2009). doi:http://doi.acm.org/10.1145/1594977.1592576
4. Guo, C., Wu, H., Tan, K., Shi, L., Zhang, Y., Lu, S.: DCell: a scalable and fault-tolerant network structure for data centers. ACM SIGCOMM Comput. Commun. Rev. 38(4), 75–86 (2008)
5. Guo, C., Lu, G., Li, D., Wu, H., Zhang, X., Shi, Y., Tian, C., Zhang, Y., Lu, S.: BCube: a high performance, server-centric network architecture for modular data centers. ACM SIGCOMM Comput. Commun. Rev. 39(4), 63–74 (2009)
6. Niranjan Mysore, R., Pamboris, A., Farrington, N., Huang, N., Miri, P., Radhakrishnan, S., Subramanya, V., Vahdat, A.: Portland: a scalable fault-tolerant layer 2 data center network fabric. ACM SIGCOMM Comput. Commun. Rev. 39(4), 39–50 (2009)

Chapter 5
Performance Enhancement

5.1 Introduction

We discussed basic routing schemes for various architectures in Chap. 4. These schemes enable basic point-to-point communication.

However, as described in Sect. 3.5.4, most data center networks offer hardware redundancy. Hardware redundancy is not only useful in dealing with potential failures, but can also be exploited to enhance network performance when there are no failures. In this section, we discuss several approaches proposed by researchers to exploit the redundant hardware to enhance network performance under non-faulty conditions; using hardware redundancy to deal with failures will be discussed in Chap. 6.

5.2 Centralized Flow Scheduling in Fat-Tree Architecture of Al-Fares et al. [1]

Al-Fares et al. propose a centralized flow scheduler for their fat-tree architecture [1]. Large flows with relatively long durations are the main concern. It is assumed in this scheme that at most one large flow originates from one host at a time. Only two kinds of components, edge switches and the central flow scheduler take part in the management of flows. An edge switch assigns a new flow to its least loaded port by local judgment. However, it will keep tracking the growth of these flows. Once a flow grows above a threshold, the edge switch will send a notification to the central scheduler with information about the large flows. The central scheduler is responsible for reassigning paths to large flows. It will keep track of all the large flows in the network. When it gets notification of a new inter-pod large flow, it scans the core switches to find one that is on a path without reserved links. On finding such a path, the central scheduler will mark all its links as reserved, and inform the relevant switches on the path. For intra-pod large flows, the central scheduler only needs to carry out a scan of the aggregation switches in that pod to find a path without reserved link. When a flow is not updated for a given period of time, the central scheduler will decide that the flow is no longer active, and will clear the reservations. During the execution of the scanning and path assignment procedure described above, the originating edge switch of the large flow will continue sending packets; in other words, the flow is not halted during the scheduler's actions.

5.3 Hedera's Flow Scheduling for Fat-Tree

A centralized flow scheduling system is also proposed in the Hedera architecture [2]. Large flows are detected at edge switches, and path selection is based on the estimated demand of large flows.

Network flows are limited by the bandwidth between the source and destination, which includes NIC capacity. The target of the centralized scheduler is to divide and assign the available bandwidth equally among existing flows. Hedera uses a $N \times N$ matrix to estimate the demand of all flows, where N is the total number of servers. An element $[i, j]$ contains three values: the number of flows from server i to server j, the estimated demand of each of the flows, and a flag marking whether the demand of a flow has converged (i.e., fully utilizing the available bandwidth without limiting other flows). It is assumed that each of the multiple flows from the same source to the same destination has the same demand, and all the demands are expressed in percentages, assuming that the path capacities between all pairs of servers are the same. The original values consider the capacity of sources only. The idea of the algorithm is to repeat iterations of increasing the flow capacities from the sources and decreasing exceeded capacities at the receivers until all demands converge.

A simulated annealing algorithm is used in the scheduler [2]. The elements of the algorithm are:

- State s: a set of mappings from destination servers to core switches. Each server is assigned to a core switch through which the incoming flows of the server are transmitted. By limiting each server to a core switch, the cost of searching paths can be reduced significantly, although the result may not be globally optimal.
- Energy function E: the total exceeded capacity of the current state.
- Temperature T: number of remaining iterations.
- Neighbor generator function: swaps the assigned core switches of a pair of servers in the current state s to get to a neighboring state s_n.
- Acceptance probability P: transition from current state s to a neighboring state s_n. P is a function of T, E, E_n, where

$$P\left(E_n, E, T\right) = \begin{cases} 1, E_n < E \\ e^{c(E-E_n)/T}, E_n \geq E \end{cases}$$

c is a constant. A suggested value for c is $0.5 \times T_0$ for a 16-host cluster, and $1,000 \times T_0$ for larger data centers [2].

In each iteration of the simulated annealing algorithm the system changes its state to state s_n with probability P, and a decrease in temperature T. When T drops to 0, the algorithm terminates. Allowing transitions to a state with higher energy can avoid local minima.

5.4 Random Traffic Spreading of VL2

VL2 [3] makes use of two techniques to offer hot-spot-free performance: Valiant Load Balancing (VLB) and Equal Cost MultiPath (ECMP). VL2 implements VLB by sending flows through a randomly-chosen intermediate switch, and ECMP by distributing flows across paths with equal cost. Path selection of VLB and ECMP is random, which means no centralized coordination and traffic engineering is needed. However, if there are large coexisting flows, random distribution of flows may lead to persistent congestion on certain links while leaving other links idle. According to the results shown in [3], this does not become a real problem for architectures with tree topologies.

5.5 BCube Source Routing

BCube Source Routing (BSR) [4] is a source routing protocol designed for the BCube architecture. Recall from Sect. 4.3.2.2, that in BCubeRouting only single paths are used to transfer the packets between two servers, even though there can be $k+1$ parallel paths between them. Thus BCubeRouting wastes most of the BCube network capacity. To fully utilize the capacity of a BCube network and enable automatic load-balancing, BSR is proposed in [4]. In BSR, the network traffic is divided into flows.

BSR uses a reactive mechanism. When a new flow is initiated, the source node sends probe packets along multiple parallel paths towards the destination. Intermediate nodes reply to the probe packets with necessary information about link state. With the information returned from intermediate nodes, the source can make routing decisions. When the source receives the responses, it computes a metric from the received information, and selects the best path. While this path selection procedure is being executed, the source node will not hold the packets of the flow in a buffer; instead, it chooses an initial path from the parallel set of paths and starts packet transmission. After the path selection procedure is completed, the source node will switch the flow to the newly selected better path. This may cause temporary out-of-sequence delivery at the destination.

The links used by a flow may fail during transmission. To address this problem, the source node sends probes periodically to detect network failures. When an intermediate node finds the next hop of a packet is unreachable, it will send a path unavailable message to the source node. The source will immediately switch the flow to the next best path in the parallel set of paths (to avoid packet buffering). When the periodic probing timer expires, the source will initiate a new path selection procedure.

Multiple concurrent flows between two nodes may be assigned to the same path, which causes both competition for resources and simultaneous execution of the periodic probing procedure. To avoid such simultaneous probes, random values are used for the periodic path selection timer.

5.6   Traffic De-multiplexing in c-Through

Hybrid networks such as c-Through [6] face a special problem: how to assign traffic to two different types of networks, electrical packet-switched and optical circuit-switched. In c-Through, the problem is solved by with VLANs. The ToR switches are assigned to two different VLANs that are logically separated. This distinguishes the optical network from the electrical network. Since the optical VLAN may be reconfigured frequently according to changing traffic demands, spanning tree protocol is disabled in optical VLANs due to its long convergence time. A management daemon runs on each server to distribute traffic to the optical and electrical networks. The path taken by each packet is determined by its destination. In practice, to make more efficient use of the optical network, optically connected destinations will have higher transmission priority than electrically connected destinations.

5.7   Summary of Performance Enhancement Schemes

The schemes discussed above use the redundant hardware available in data center networks to improve performance. In this section we summarize these schemes in order to develop guidelines for performance enhancement. Table 5.1 shows a comparison of the different performance enhancement schemes.

Table 5.1 Performance enhancement schemes

	Fat-tree [1]	BCube [4]	Hedera [2]	VL2 [3]
Flow scheduling algorithm	Reassign paths for large flows	Metric-based routing	Simulated annealing	Random path selection
Decision maker in path selection	Central scheduler	Source node	Edge switches	Source node

5.7.1 Use of Multiple Paths for Performance Enhancement

The redundant hardware in data center networks enables the availability of multiple paths between same source-destination pairs. Schemes that make better use of idle components, and avoid overloaded components will improve overall performance. Some conditions must be met to enable the use of multiple paths:

1. There must be hardware redundancy. Topologies such as the basic tree have no hardware redundancy. There is only one deterministic path between two servers, and hence performance enhancement schemes that rely on multiple paths are not feasible.
2. Route selection schemes should consider fairness when assigning data flows to paths, i.e., data traffic should be dispersed, and every path is not necessarily the "best" path. (The interpretation of the term "best" depends on the metric used to compare paths; for example, a "best" path can be a path with the smallest latency, or the largest bandwidth.)
3. Paths may overlap at some links/nodes.

It should be noted that in data centers with hardware redundancy, even if source routing is not used, there are other ways to make use of the idle redundant links: nodes can detour packets according to local information about network conditions. This is also referred to as detour routing. It offers a more flexible way to make use of idle links. The disadvantage of detour routing is that there must be mechanisms to avoid possible loops.

Multiple paths to a destination may not be used concurrently. Instead, the multiple paths may simply be stored at the source node, and consulted only if the current path to a destination becomes unavailable. However, the ability to use multiple paths concurrently is helpful to disperse traffic more evenly. Such use requires careful design to avoid out-of-sequence delivery of packets. Flow scheduling schemes are designed to address this problem.

5.7.2 Flow Scheduling

Identifying flows in the traffic allows for routing schemes that can better achieve fairness and efficiency. However, today's routers are typically not configured to execute automatic flow identification. Flow classification is available, and firewall filters can be used to have routers choose paths other than the path indicated in the default IP routing tables, but such mechanisms are used sparingly. Most packets are forwarded based on the default routing table entries.

Flow scheduling is meaningful only when there are multiple optional paths, so that flows can be distributed among the different paths. There are two approaches to directing a flow: decide the whole path of the flow at the source, or make routing decisions at intermediate nodes. The latter approach needs every intermediate node to identify a flow, while in the former approach intermediate nodes need to only extract routing information from a packet header to determine the next hop on the path.

Flow scheduling solves the out-of-sequence delivery problem introduced by the use of concurrent multiple paths. Since data packets in the same flow are transferred sequentially on a given path, out-of-order delivery is avoided.

Flow scheduling, if executed online (i.e., on live traffic), needs information to be collected from the first few packets of a flow, while at the same time these packets may have been sent through different paths. According to recent findings [5] about flows in data centers, more than 80% of flows whose durations are shorter than 10 s, while only fewer than 0.1% of flows last more than 2,000 s. On the other hand, 50% of data are in flows whose durations are shorter than 25 s. The main challenges in scheduling flows include:

1. It is difficult to predict whether or not a flow will be a large flow; if two large flows are routed to the same path, the usage on multiple paths could become unbalanced.
2. The first packets of a flow may arrive out of sequence. Given that most flows last a short time, and most data are in these short flows, the out-of-sequence delivery problem cannot be solved for these short flows.

Thus, though flow scheduling offers benefits such as balanced link loads and improved performance, it is challenging to implement.

References

1. Al-Fares, M., Loukissas, A., Vahdat, A.: A scalable, commodity data center network architecture. In: Proceedings of the ACM SIGCOMM 2008 Conference on Data communication, Seattle, pp. 63–74. ACM (2008)
2. Al-Fares, M., Radhakrishnan, S., Raghavan, B., Huang, N., Vahdat, A.: Hedera: dynamic flow scheduling for data center networks. In: Proceedings of the 7th USENIX Conference on Networked Systems Design and Implementation, San Jose, p. 19. USENIX Association (2010)
3. Greenberg, A., Hamilton, J.R., Jain, N., Kandula, S., Kim, C., Lahiri, P., Maltz, D.A., Patel, P., Sengupta, S.: VL2: a scalable and flexible data center network. SIGCOMM Comput. Commun. Rev. 39(4), 51–62 (2009). doi:http://doi.acm.org/10.1145/1594977.1592576
4. Guo, C., Lu, G., Li, D., Wu, H., Zhang, X., Shi, Y., Tian, C., Zhang, Y., Lu, S.: BCube: a high performance, server-centric network architecture for modular data centers. ACM SIGCOMM Comput. Commun. Rev. 39(4), 63–74 (2009)
5. Kandula, S., Sengupta, S., Greenberg, A., Patel, P., Chaiken, R.: The nature of data center traffic: measurements & analysis. In: Proceedings of the 9th ACM SIGCOMM Conference on Internet Measurement Conference, Chicago, pp. 202–208. ACM (2009)
6. Wang, G., Andersen, D., Kaminsky, M., Papagiannaki, K., Ng, T., Kozuch, M., Ryan, M.: c-Through: part-time optics in data centers. In: ACM SIGCOMM Computer Communication Review, vol. 40, pp. 327–338. ACM, New York, NY, USA (2010)

Chapter 6
Fault-Tolerant Routing

6.1 Introduction

Considering the large number of servers and switches used in a data center, it is unlikely that the system can keep running without failures throughout its lifetime. A data center network must be designed to be able to recover automatically from common failures, and maintain most of its performance in the interim. At the hardware level, most data center topologies employ redundant links to ensure connectivity in case of hardware failures. Moreover, the routing mechanisms running above make use of the redundancy offered by the hardware to recover from communication failures. Since we have discussed hardware redundancy in Sect. 3.5.4, we will turn to the software mechanisms, i.e., the fault-tolerant routing protocols used by different data center networks.

6.2 Failure Models

Failures in data centers can occur in different components. Classifying the failures into *failure models* helps to figure out the characteristics of different failures and methods to detect and deal with certain types of failures. Failures can be classified based on the following aspects.

6.2.1 Failure Type

This attribute identifies the level at which components are diagnosed as having failed [3]. Either the whole component fails, or one of the links connected to it fails. The former is referred to as *node failure*, and the latter as *link failure*.

For components that have only one link, such as the servers in a fat-tree, node failure and link failure have the same effect. On the other hand, a node failure of a node with multiple links can be considered as concurrent failures on all its links. A mechanism that can deal with link failures can also deal with node failures. However, this does not preclude the need for distinguishing between node failures and link failures. For example, marking a node as failed means that the routing protocol can choose to detour around that node without checking its links one by one, thus saving significant time.

It should be noted that link failures and node failures are not necessarily hardware failures. The real meaning of a link failure or node failure is that a link or all the links connected to a node are no longer available to carry data from an application's point of view.

6.2.2   Failure Region

A *failure region* is a useful concept when multiple related components are failed. For example, in a data center constructed of racks, power failure of a rack will disconnect all the components in the rack. If a routing protocol can detect rack failures, it can bypass the whole rack. Since positions of components depend on the physical architecture and logical architecture of a data center, failure regions can vary according to the architecture.

6.2.3   Failure Neighborhood

This attribute reflects the extent of dissemination of failure information. Based on the distance from the failed node, there are three different modes: *global*, in which every node in the network has information about all failed nodes; *local*, in which only the adjacent nodes of a failed node are aware of its status; *k-neighborhood*, information about failed nodes is disseminated to nodes within a distance of *k* hops. The global mode can lead to better route choices, but it needs more storage and periodic global synchronization. On the other hand, routing exchanges are simpler and faster in the local mode, but routes could be longer. The *k*-neighborhood mode offers a balance between the two extremes.

6.2.4   Failure Mode

This attribute is determined by the duration of the failure. A *static* failure is one that was present from the instant the system was initiated, while a *dynamic* failure is one that occurs during the operation of a system. Both kinds of failures are considered to

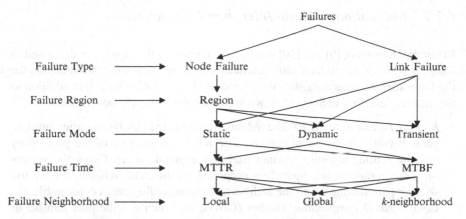

Fig. 6.1 A taxonomy for faults

be *permanent*. In contrast, a failure can be *transient*. Dynamic failures and transient failures pose challenges in designing a robust routing protocol.

6.2.5 Failure Time

This attribute characterizes the frequency of component failures. Two parameters, the *mean time between failures (MTBF)* and *mean time to repair (MTTR)*, are used to evaluate the possibility of a failure and the time needed to recover from it. For example, if MTBF and MTTR are close for one kind of component, it is likely that there will be multiple concurrent failed components in the data center. Prior work on MTTR [4] shows that 95% of failures are resolved in 10 min, 98% in less than an hour, 99.6% in less than a day, but 0.09% failures last for more than 10 days. These percentages may vary in data centers using different hardware, but serve as a reference for the distribution of MTTR.

6.2.6 Taxonomy of Faults

Figure 6.1 shows a taxonomy for the failure models discussed above. One failure can be classified into multiple models. Ideally, routing algorithms should account for all models that can be constructed with various combinations of failure characteristics.

6.2.7   Evaluation of Fault-Tolerance Characteristics

In our previous work [9] and [10] we evaluated some of the topologies discussed in Chap. 3, with respect to their fault-tolerance characteristics. Simulation results for the fault-tolerance characteristics were generated using DCNSim [8]. In addition to the metrics described in Sect. 3.5.3, we introduce the following new metrics:

- *Routing Failure Rate (RFR)*. The redundant hardware in DCNs can offer multiple paths between each pair of servers. The *RFR* is a measure of the probability that the routing algorithm cannot find any available route. Given the variety of fault-tolerant routing algorithms proposed for different architectures, we use shortest path routing for all the architectures to make the results comparable.
- *Component Decomposition Number (CDN)*. This metric is the total number of connected components in the presence of failures. Assume that the connected components of a DCN with failures are labeled as C_1, C_2, $\ldots$, C_t, then t is the *CDN* of the network. Smaller CDN indicates that the network is separated to fewer connected components.
- *Smallest/Largest Component Size (SCS/LCS)*. The size of a connected component is defined as the number of servers inside it. For applications that require a certain number of servers to collaborate, it is important to find a connected component of sufficient size. We use *SCS/LCS* to evaluate the size of the connected components, while $SCS = \min_{1 \le i \le t} |C_i|$, and $LCS = \max_{1 \le i \le t} |C_i|$. $|C_i|$ denotes the size of the i-th connected component.

RFR, along with *ABT* and *APL* (which were defined in Sect. 3.5.3), are categorized as *connection-oriented metrics*. These metrics evaluate connectivity between servers as determined by the routing algorithms.

On the other hand, *CDN* and *SCS/LCS* are categorized as *network size-oriented metrics*. These metrics characterize the sizes of the (still) connected components into which the original topology is divided due to its failures. Although it is impossible for a server to communicate with servers in other connected components, the servers in the same connected component can still communicate with each other. Such connected components can be regarded as DCNs of smaller size, and can satisfy certain application requirements.

6.2.7.1   Connection-Oriented Metrics Under Link Failures

Simulation results are shown in Fig. 6.2 for the connection-oriented metrics, assuming uniformly distributed random link failures. The number of servers, N, are marked in the figures. Figure 6.2a shows a comparison of ABT. Fat-tree and BCube achieve very high ABT, but the ABT of BCube degrades more gracefully because of high redundancy. DCell has relatively lower ABT, however the cost of switching and wiring is also lower.

The average path length of the fat-tree does not change significantly because all the backup routes are of the same length, and randomly distributed failures does not

Fig. 6.2 Connection-oriented metrics under link failures. (**a**) ABT. (**b**) APL. (**c**) RFR

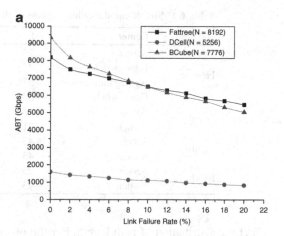

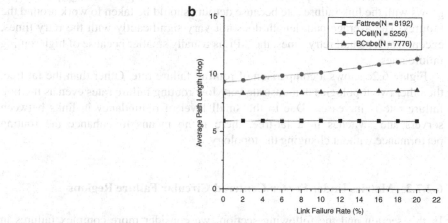

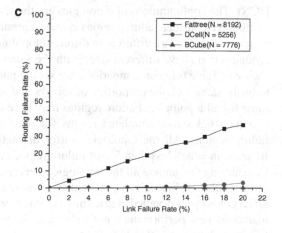

Table 6.1 Size of circular failure regions of different topologies

Topology	Center	Links	Servers	Switches
Fat-tree	Server	24	1	1
	Edge switch	288	12	13
	Aggregation switch	576	0	25
	Core switch	576	0	25
$DCell_3$	Server	13	3	1
	Switch	21	7	1
$BCube_3$	Server	45	1	3
	Switch	45	15	1
$BCube_5$	Server	25	1	5
	Switch	25	5	1

affect the distribution of path length. For the other topologies, average path length grows with the link failure rate because detours should be taken to work around the failures. The average path length does not vary significantly with the retry times, except that for small retry times, the APL is usually smaller because of high routing failure rates.

Figure 6.2c shows a comparison of routing failure rate. Other than the fat-tree, the other two topologies demonstrate very low routing failure rates even as the link failure rate is increased. Due to the small level of redundancy in links between servers and switches in a fat-tree, there is no means to enhance the routing performance without changing the topology.

6.2.7.2   Metrics Under Server-Centered Circular Failure Regions

In this section and the following section, we consider more complex failures in DCNs. The configurations of topologies are the same as in Table 3.4.

First we consider failure regions centered around a server. Since the connection patterns of servers are different in different topologies, the circular regions centered around servers have different sizes with the same radius, which are computed and shown in Table 6.1. For comparison, we set the failure rate for such circular regions to be the same, i.e., the proportion of servers that are centers of failure regions is the same for all topologies. Failure regions may have overlaps.

Figure 6.3 shows simulated results for metrics under server-centered circular failure regions. All the topologies suffer dramatic drops in ABT, especially for $BCube_3$, in which less than 5% of failure regions can bring the ABT down by half. Considering that among all the topologies, server-centered failure region causes the most number of link failures in $BCube_3$, it is reasonable that such failure regions affect connection-oriented metrics most in this topology. $BCube_5$, on the other hand, maintains best performance, not only because the number of link failures in its failure regions is smaller, but also because the large number of its links makes the proportion of failed links even smaller.

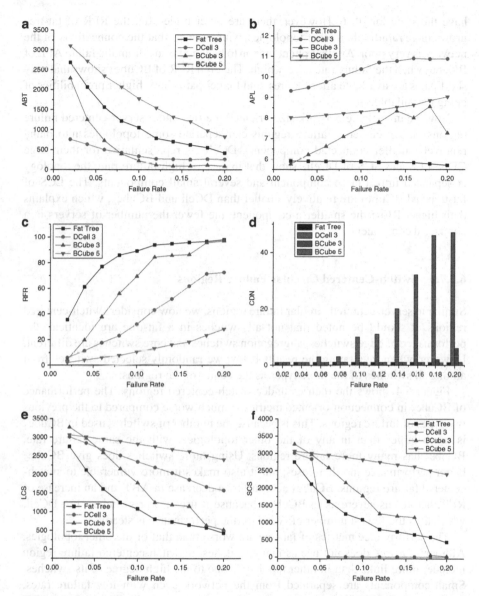

Fig. 6.3 Metrics under server-centered failure regions. (**a**) ABT. (**b**) APL. (**c**) RFR. (**d**) CDN. (**e**) LCS. (**f**) SCS

Corresponding to the decrease in ABT, there is also notable increase in APL for some of the topologies. One exception is the fat-tree, whose APL remains almost steady with increasing failure rate. This is because all the backup paths in fat-tree have the same path length, e.g., all the paths between two servers in different pods have the same length 6. However, there are other trade-offs: the RFR of fat-tree grows more rapidly than other topologies, which means that the connectivity of the network is very poor. Another phenomenon to be noted is the dramatic fall in APL of $BCube_3$ when the failure rate is very high. The high RFR of $BCube_3$ shows that only short paths are available after failures, and longer paths have higher probabilities of being unavailable.

Now we investigate network size-oriented metrics under server-centered failure regions. Server-centered failure regions can separate some topologies into many relatively smaller connected components. $DCell_3$ is representative for their large CDN. LCS and SCS of $DCell_3$ show that in case of high failure rate, the topology is separated into a large component and several small components. The LCS of fat-tree and $BCube_3$ are relatively smaller than DCell and $BCube_5$, which explains their higher RFR: the smaller a component, the fewer the number of servers in a connected component.

6.2.7.3   Switch-Centered Circular Failure Regions

Similar to server-centered circular failure regions, we now consider switch-centered regions. It should be noted that not all switches in a fat-tree are identical; the performance of edge switches, aggregation switches and core switches are different. However, when generating the results below we randomly selected switches from among all switches of the three types as the center of the regions.

Figure 6.4 shows the metrics under switch-centered regions. The performance of $BCube_5$ in connection-oriented metrics is much worse compared to the previous two types of failure regions. This is because the number of switches used in $BCube_5$ is much larger than in any of the other topologies; with the same failure rate, $BCube_5$ has many more failure regions. Using more switches does give $BCube_5$ better performance in other cases, but it also makes it more vulnerable to switch-centered failure regions. $BCube_3$ also suffers a decrease in ABT and an increase of RFR, but not as severe as in $BCube_5$, because it uses fewer switches. For DCell, which uses the fewest number of switches, the performance is steady.

The performance metrics of fat-tree are worse than that of the other topologies. Although fat-tree does not use as many switches, a switch-centered failure region includes more links than in other topologies due to the high degree of its switches. Small components are separated from the network even with low failure rates, as shown in Fig. 6.4f. It should be noted that there is a sudden increase in ABT of fat-tree before it shows a continued decreasing trend, which corresponds to an increase in RFR and CDN. This indicates that the network is separated in to many components, disconnecting flows between these components. With enough LCS which is shown in Fig. 6.4e, ABT is still acceptable by summing up all the inner-component traffic. This in fact means that most traffic are limited to servers within the small components. Fat-tree is unsuitable to support data center applications that require a minimum number of connected servers.

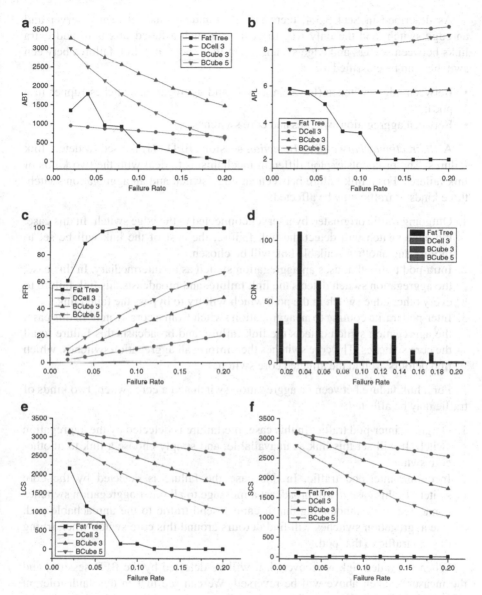

Fig. 6.4 Metrics under switch-centered fault regions. (**a**) ABT. (**b**) APL. (**c**) RFR. (**d**) CDN. (**e**) LCS. (**f**) SCS

6.3 Link Failure Response in Fat-Tree

Al-Fares et al. [1] proposed a failure broadcast protocol to detect link failures in their fat-tree architecture. The protocol handles only simple link failures; the other failure models described in Sect. 6.2 are not mentioned.

As described in Sect. 3.5.4, there is no redundant link between a server and an edge switch, and the only way to deal with such a failed link is to add extra links between servers and edge switches. On the other hand, link failures between switches can be classified as:

- Between edge switches (lower in a pod) and aggregation switches (upper in a pod);
- Between aggregation switches and core switches.

A *Bidirectional Forwarding Detection* session (BFD [7]) is used to detect link failures. The protocols exploit different mechanisms to deal with the two kinds of link failures. For a link failure between an edge switch and an aggregation switch, three kinds of traffic may be affected:

1. Outgoing traffic originated by a server connected to the edge switch. In this case, the edge switch will detect the link failure, the cost of the link will be set to infinity and another available link will be chosen.
2. Intra-pod traffic that uses an aggregation switch as an intermediary. In this case, the aggregation switch detects the link failure, and broadcasts this link failure to every other edge switch in the pod, which will try to bypass the failed link.
3. Inter-pod traffic coming to an aggregation switch from a core switch. In this case, the aggregation switch detects the link failure, and broadcasts this failure to all the core switches. The core switches then inform all aggregation switches, which find detours around the affected core switch.

For a link failure between an aggregation switch and a core switch, two kinds of traffic may be affected:

1. Outgoing inter-pod traffic. In this case, the failure is detected by the aggregation switch. It will set this link as unavailable, and simply choose a link to another core switch.
2. Incoming inter-pod traffic. In this case the failure is detected by the core switch. In this case it will broadcast a message to all other aggregation switches connected to it, announcing that it cannot send traffic to the unreachable pod. The aggregation switches will find detours around this core switch when trying to send traffic to that pod.

When the failed link is recovered, it will be detected by the BFD session, and the measures taken above will be reversed. We can see that in this fault-tolerant scheme the failure information is sent to nodes at most two hops away, i.e., it uses a 2-neighborhood mode.

6.4 Fault-Tolerant Routing in PortLand and Hedera

Mysore et al. [11] proposed a fault-tolerant routing scheme for PortLand, which uses link failures to represent all possible failures. Since PortLand exploits a centralized fabric manager to maintain the entire network, it is more effective in collecting and spreading fault messages. The periodic LDM mentioned in Sect. 4.2.2.4 also serves as a probe of link failures. If a node does not receive an LDM for a certain period of time, it will assume that a link failure has occurred and send the information to the fabric manager. The fabric manager uses a *fault matrix* to store the status of each link in the whole network. On receiving a fault message, the fabric manager updates the fault matrix, and informs all the affected nodes of this link failure. These nodes will use the new information to reroute packets. On recovery from a link failure, the procedure is similar except that the information is a recovery message instead of a fault message. With the fabric manager, the extent of the failure information can be controlled easily. Besides, by using multicast instead of broadcast, the cost of spreading failure messages can be reduced.

Hedera [2], whose implementation augments PortLand routing and fault-tolerant protocols, has the same mechanisms for dealing with failures.

6.5 DCell Fault-Tolerant Routing

DCell Fault-tolerant Routing (DFR) uses DCellRouting and DCellBroadcast to enable fault-tolerant routing in DCell. It defines three types of failures: server failure, rack failure and link failure, and it uses three techniques, local reroute, local link-state, and jump-up, to deal with these types of failures respectively.

- Local Reroute: bypass a link failure using only local decisions. When a link fails, local reroute will be executed at one of the two ends of the failed link. If the failed link (n_1, n_2) is a level-l link (i.e., the two ends of the link is in the same DCell$_l$ but in different DCell$_{k-1}$s), n_1 will choose a node in another DCell$_{k-1}$ that does not contain n_2 as its *proxy*. As long as at least one link from n_1's DCell$_{k-1}$ to each of the other DCell$_{k-1}$s is available, such a proxy can always be found. After the proxy is selected, DCellRouting will be executed on the proxy to find a path to the destination. Local reroute may be used to transmit packets to the other end of a failed link, but if it is a failed node, local reroute cannot bypass it.
- Local Link-state: use link-state information gathered with DCellBroadcast to bypass failed nodes. By means of DCellBroadcast, a node in a DCell$_b$ (where b defines the range within which a broadcast packet is delivered, which generally include a rack of switches and its servers) learns the status of all the links connecting its own DCell$_b$ and other DCell$_b$s. In this scheme, a node will compute the route inside its DCell$_b$ local using link-state information and Dijkstra's algorithm. When a link fails, the node will invoke local-reroute to find a proxy. Since the decision is made upon receiving link-state information for the whole DCell$_b$, irrespective of whether the link failure implies a potential node failure, the failed node can be bypassed.
- Jump-up: a scheme to address rack failure, in which a whole DCell$_b$ fails. In local link-state, when a whole DCell$_b$ fails, the packet will be re-routed endlessly round DCell$_b$. To address such a failure, the jump-up scheme is introduced. When

a link failure is detected, and the first attempt to re-route to another path in the DCell$_b$ is unsuccessful, it will be assumed that the whole DCell$_b$, i.e., the whole rack is down. Then the receiver of the re-routed packet will "jump-up" to a higher level DCell to bypass the whole failed DCell$_b$. Besides, since it is impossible to reach a node inside a failed DCell$_b$, a time-to-live (TTL) field is added into every packet header to avoid endless re-routing.

By using all the techniques above, DFR can handle link failures, node failures and rack failures.

6.6 Fault-Tolerant Routing in BCube

As discussed in Sect. 5.5, BCube can detect link failures by sending periodic probes, and use backup paths when failure occurs.

6.7 Summary of Fault-Tolerant Routing Algorithms

We discussed some fault-tolerant routing techniques in this chapter. From a comparison shown in Table 6.2, we offer some guidelines for designing fault-tolerant routing algorithms. Some of these guidelines are drawn from the performance enhancement techniques presented in Chap. 5, while others are presented in this chapter.

1. All algorithms have their own mechanisms to detect link failures. Since node failures can be recognized as a set of link failures, the mechanisms developed to cope with link failures can be applied to node failures or even rack failures as these failures are equivalent to multiple failed links.
2. The MTTR and MTBF of different components should be considered when designing fault-tolerant mechanisms, e.g., when deciding the frequency for executing failure detection processes.
3. Spreading failure information can reduce the time overhead incurred in waiting for a timeout. However, such dissemination may involve non-adjacent nodes, and updates could occupy a significant fraction of bandwidth.

Some fault-tolerant techniques such as maintaining back-up routes in routing tables can serve to enhance performance when there are no faults. It is meaningful to consider both aspects, performance enhancement and fault-tolerance, together when designing routing algorithms.

Table 6.2 Fault-tolerant techniques

	Fat-tree [1]	DCell [5]	BCube [6]	PortLand [11] & Hedera [2]
Failure types	Link failure	Link failure/Server failure/Rack failure	Link failure	Link failure
Failure detecting mechanisms	BFD (Periodic probes)	DCellBroadcast (Reactive probes)	Periodic probes	LDM (Periodic probes)
Area of spreading failure information	Edge/Aggregation/Core	Local/Rack/Inter-rack	Source	All nodes on the path
Decision maker	Intermediate	Intermediate	Source	Fabric manager

References

1. Al-Fares, M., Loukissas, A., Vahdat, A.: A scalable, commodity data center network architecture. In: Proceedings of the ACM SIGCOMM 2008 Conference on Data Communication, Seattle, pp. 63–74. ACM (2008)
2. Al-Fares, M., Radhakrishnan, S., Raghavan, B., Huang, N., Vahdat, A.: Hedera: dynamic flow scheduling for data center networks. In: Proceedings of the 7th USENIX Conference on Networked Systems Design and Implementation, San Jose, p. 19. USENIX Association (2010)
3. Duato, J., Yalamanchili, S., Ni, L.: Interconnection Networks: An Engineering Approach. Morgan Kaufmann, San Francisco (2003)
4. Greenberg, A., Hamilton, J.R., Jain, N., Kandula, S., Kim, C., Lahiri, P., Maltz, D.A., Patel, P., Sengupta, S.: VL2: a scalable and flexible data center network. SIGCOMM Comput. Commun. Rev. **39**(4), 51–62 (2009). doi:http://doi.acm.org/10.1145/1594977.1592576
5. Guo, C., Wu, H., Tan, K., Shi, L., Zhang, Y., Lu, S.: DCell: a scalable and fault-tolerant network structure for data centers. ACM SIGCOMM Comput. Commun. Rev. **38**(4), 75–86 (2008)
6. Guo, C., Lu, G., Li, D., Wu, H., Zhang, X., Shi, Y., Tian, C., Zhang, Y., Lu, S.: BCube: a high performance, server-centric network architecture for modular data centers. ACM SIGCOMM Comput. Commun. Rev. **39**(4), 63–74 (2009)
7. Katz, D., Ward, D.: Bfd for ipv4 and ipv6 (single hop). draft-ietf-bfd-v4v6-1hop-09 (work in progress) (2009)
8. Liu, Y., Muppala, J.: DCNSim: a data center network simulator. In: 33rd International Conference on Distributed Computing Systems Workshops (ICDCSW), Philadelphia, 2013, pp. 1–6. IEEE (2013)
9. Liu, Y., Muppala, J.: Fault-tolerance characteristics of data center network topologies using fault regions. In: IEEE/IFIP 43rd International Conference on Dependable Systems and Networks Workshops (DSN-W), Budapest, 2013, pp. 1–6. IEEE (2013)
10. Liu, Y., Lin, D., Muppala, J., Hamdi, M.: A study of fault-tolerance characteristics of data center networks. In: IEEE/IFIP 42nd International Conference on Dependable Systems and Networks Workshops (DSN-W), Boston, 2012, pp. 1–6. IEEE (2012)
11. Niranjan Mysore, R., Pamboris, A., Farrington, N., Huang, N., Miri, P., Radhakrishnan, S., Subramanya, V., Vahdat, A.: Portland: a scalable fault-tolerant layer 2 data center network fabric. ACM SIGCOMM Comput. Commun. Rev. **39**(4), 39–50 (2009)

Chapter 7
Conclusions

In this book, we examined data center networks in detail, focusing on some representative topologies of data center networks. First, we reviewed the current state-of-the-art for data center topologies. Then, we introduced several of the newly proposed topologies that have appeared in research and academic literature. We categorized these topologies according to their features. We analyzed several of the properties of these topologies, in terms of their scalability, performance and hardware redundancy and presented some performance results by comparing various metrics.

We explored the routing algorithms designed for various architectures, and the techniques for enhancing their performance taking advantage of available hardware redundancy. We examined the fault-tolerance characteristics of the architectures. Besides qualitative analysis, we also presented some quantitative results computed through simulations to compare the fault-tolerant characteristics of the architectures.

With this endeavour, we hope that the stage has been set to examine various DCN architectures using a standard set of metrics. We also expect that the detailed categorization and characterization of various aspects of the architectures will enable future researchers to develop new architectures and propose mechanisms to improve performance and dependability characteristics. Any architecture has to strike a balance between performance, cost and sustainability.

Y. Liu et al., *Data Center Networks*, SpringerBriefs in Computer Science, DOI 10.1007/978-3-319-01949-9_7, © The Author(s) 2013

Index

Y. Liu et al., *Data Center Networks*, SpringerBriefs in Computer Science,
DOI 10.1007/978-3-319-01949-9, © The Author(s) 2013